THE ILLUSTRATED HISTORY OF
PISTOLS
REVOLVERS
and SUBMACHINE GUNS

THE ILLUSTRATED HISTORY OF
PISTOLS
REVOLVERS
and SUBMACHINE GUNS

A fascinating guide to small arms development,
covering the early history through to the modern age

WILL FOWLER, ANTHONY NORTH AND CHARLES STRONGE

LORENZ BOOKS

This edition is published by Lorenz Books,
an imprint of Anness Publishing Ltd,
Blaby Road, Wigston, Leicestershire LE18 4SE; info@anness.com

www.lorenzbooks.com; www.annesspublishing.com

If you like the images in this book and would like to investigate using them for publishing, promotions
or advertising, please visit our website www.practicalpictures.com for more information.

Publisher: Joanna Lorenz
Editorial Director: Helen Sudell
Project Managers: Sarah Doughty and Rosie Gordon
Photography: Gary Ombler
Designers: Alistair Plumb and Jonathan Harley
Jacket Designer: Michael Morey
Art Director: Lisa McCormick
Production Controller: Mai Ling Collyer

Previously published as part of another title, *The World Encyclopedia of Pistols, Revolvers and Submachine Guns*

The publisher would like to thank the following for kindly supplying photos for this book: Corbis: 17t, 34b,
36, 54b, 90t, 94, 95t, 95b; Getty Images: 43, 79b, 37t; Horst Held: 49t; iStockphoto: 26b, 33b, 39t, 39tm, 65bl,
82b, 83t, 86m, 86b, 89b, 89t, 90b, 91r, 92; JupiterImages 33r, 39, Peter Newark's Military Pictures: 11t, 13b, 16b,
19b, 27b, 27t, 28bl, 29t, 31b, 31t, 37b, 38, 42, 47b, 63t, 75mr, 11b; Royal Armouries Picture Library: 15t, 17b,
18t, 20t, 21b, 21t, 22t, 24t, 25t, 26t, 33t, 35b, 41b, 44bl, 45b, 48, 53t; The Bridgeman Art Library: 6b, 23, 41t;
The Kobal Collection: 87b; The Research House: 78b, 83b, 85t, 85b; TopFoto: 10, 15b, 19tr, 28t, 29b, 40t, 47t,
52b, 63b, 68bl; Will Fowler: 58b, 60b, 62b, 64b, 66b, 67b, 70b, 71b, 72b, 76b, 51b.

All other images are commissioned. With thanks to the Royal Armouries, Leeds in England for allowing access to
their extensive collection of firearms for photography. With thanks to the following companies for supplying
images: Browning-Winchester, Glock, Heckler & Koch, Para-Ordnance, Ruger, Tangfoglio, Taurus International
Manufacturing, Inc. All commissioned pictures by Gary Ombler. All artwork by Peters & Zabransky Ltd., and
Richard Peters. With thanks to Cybershooters for their invaluable information about gun mechanics.

Every effort has been made to obtain permission to reproduce copyright material, but there may be cases where we
have been unable to trace a copyright holder. The publisher will be happy to correct any omissions in future printings.

PUBLISHER'S NOTE
Although the advice and information in this book are believed to be accurate and true at the time of going
to press, neither the authors nor the publisher can accept any legal responsibility or liability for any
errors or omissions that may have been made.

Contents

Introduction

Although the firearm has a long history, the amount of technical progress that took place in the first 500 years is relatively small. The earliest evidence of the handgun, whether in illuminated manuscripts or as archaeological artefacts, gives us clues to its development from an unwieldy vase-shaped object to a straight-barrelled version similar to the barrel of a modern firearm. This book traces the entire history of small-arms development, from the origins of gunpowder to the most up-to-date weapons of the 21st century. It discusses the problems met by soldiers and other operators over the years, and how these problems were solved by ingenious firearms designers and manufacturers.

The earliest arms development

The first part of this book, *The Early History of Arms,* is a comprehensive history of the technical development of early firearms. It opens with an exploration of the very earliest inventions from the 11th century. Always improving on previous models, various techniques were subsequently developed to ignite and fire prototype guns in the most efficient way possible, while allowing the user to hold the weapon by a wooden stock (the handle). While early guns were often ignited by a burning tinder or a red-hot iron, the serpentine system allowed an arm (the holding "serpentine") a match to ignite the powder in the touch-hole. This developed into the matchlock system, whereby the arm was held clear of the powder by a spring. To fire the gun, the spring was released by means of the action of the trigger and the powder in the priming pan was lit. The wheel-lock worked very much like a flint lighter, whereby a wheelspring mechanism helped the arm to move to ignite the powder in the pan by means of friction. The 17th-century flintlock developed this idea further and a flint was struck against steel to produce a spark. The invention of the flintlock was so successful that it formed the basis of all military, naval and personal firearms for the best part of 200 years.

The percussion system was an ingenious development that saw the end of the flintlock, first by replacing the traditional locks (flintlocks or wheel-locks) on existing weapons and then by being fitted to new firearms. The percussion cap was a system using a

BELOW A 17th-century German flintlock pistol. The flintlock rapidly replaced earlier firearm-ignition. It continued to be in common use for over 200 years before it was replaced by the percussion cap.

LEFT Handguns in use in a 15th-century siege. As technology improved, guns and cannons took over from traditional medieval weapons such as crossbows and combat became more impersonal and mechanical.

cap containing fulminate of mercury, a chemical compound which exploded when struck, that was attached to the barrel of the gun by a simple tube. The subsequent experiments with percussion caps, early cartridges and breech-loading rifles (loaded from the rear rather than the front) gave rise to the modern firearm. Different trades, influenced by increasing industrialization in Europe, contributed to its manufacture, whether lock, stock or barrel.

The 20th century and beyond
The second section, *Into the Modern Age,* examines the development of firearms in the modern day. Of particular significance was the invention of the revolver by Samuel Colt, which had a critical influence on American history as it was played out in both the "Wild West" (real and mythical) and the American Civil War (1861–65).

The turn of the 20th century also saw the introduction of one of the most important handgun designs of modern history, John Moses Browning's M1900 automatic pistol; later refined to become the Colt M1911. Automatic handguns became standard issue for law enforcement officers worldwide, and original automatic pistols were so ingeniously designed that their derivates have remained in service with major military forces to the present day.

The beginning of World War II (1939–45) saw an acceleration of handgun and submachine-gun design. The British Sten gun was conjured up in the dark days of 1940, while the Soviet PPS-43 was born from the wreckage of Leningrad in 1942–43. Wartime demanded great design advances from manufacturers. For example, the German Walther P38 was cheap to make and had an innovative safety catch to prevent accidental fire. A pin indicator also told soldiers if the gun was loaded by touch, rather than relying on sight and good light conditions. The 9mm MP Erma, developed in the 1920s and used in the front

ABOVE The .45 Colt M1911, regarded as the most successful automatic pistol of all time. Adopted by the US Army in 1911, the pistol was still in service in 1985. There are many copies still in use.

line until 1944, had the first folding stock – ideal for use by paratroopers. And the M3, produced by America in 1943, was a low-cost version of the Thompson "Tommy Gun". Accessories made cleaning and operation easy, and kept the weapon reliable. Dirt did not cause the weapon to jam, which hitherto had been a real problem in trench warfare. The M3 was highly successful, and became popularly known as the "Grease Gun".

In the post-War era, further revolver, pistol and submachine-gun designs were developed in response to warfare, the needs of police forces, and the growing threat of international terrorism. In the 1950s the "Uzi" entered service with the Israeli army and was so successful that it has since been used by 90 countries. It made innovative use of existing Czech M23 and M25 submachine-gun designs. Sterling and Ingram also had considerable success with making compact submachine-guns in this period.

Silencers, pioneered by British designers in World War II, are also investigated. This essential technology has been adoped worldwide for all types of weapon, including submachine guns.

Also covered in the second part of the book is the advent of non-lethal weapons, as security forces explore the options of stunning or anaesthetizing their opponents rather than using lethal force.

Photographs of firearms collections provide a fully illustrative history of pistols, revolvers and submachine guns. They display the design of the guns, each tailored to a specific use, and the craftsmanship involved in producing them, allowing readers to appreciate the evolution of these incredible weapons.

ABOVE RIGHT The Micro Uzi is used for covert operations that require a weapon that can be easily concealed and deployed while providing an effective rate of fire and accuracy. While classified as a submachine gun, Micro Uzi is small enough to sometimes be designated as a pistol.

RIGHT A highly decorated flintlock pistol. The flintlock quickly replaced earlier firearm technologies, such as the matchlock and wheel-lock mechanisms. It continued to be in general use for over two centuries, until the percussion cap and cartridge systems came into common use.

RIGHT This was one of the most famous revolvers of the American Wild West. Designed for the US Cavalry, it was first produced in 1873 and is often known as the "Colt 45".

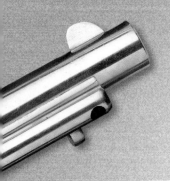

The early history of arms

This part of the book represents a colourful journey from the early discoveries of chemical compounds that would explode when lit, and the use of these powders to fire a projectile, through the development of ever more sophisticated systems for firing weapons. These included matchlock and flintlock mechanisms, and the early forms of today's firearms, which are loaded from the rear of the barrel (the breech) and use metal cartridges containing bullet, gunpowder and primer. These pages also consider the use of firearms for duelling, examine the unique nature of handguns produced by the Islamic countries of the East, and reflect on the arts of ornamentation and engraving that were to complement handgun design and development through the ages.

ABOVE From the 18th century, duels could be fought using pistols. Special sets of duelling pistols were crafted for wealthy noblemen. A pistol duel normally took place at dawn. The parties would be placed back to back with loaded weapons in hand. They would walk a set number of paces, turn to face their opponent and shoot. Many pistol duels were to first blood (such as a minor wound) rather than to death.

The first firearms

The invention of the gun was the logical consequence of the discovery of gunpowder. Once the knowledge of its powerful explosive forces had spread to Europe, probably from China via Islamic Spain, the next challenge was to contain the force and to use it to propel a missile.

The discovery of gunpowder

Gunpowder was known to the Chinese by the 11th century and the knowledge of it probably came to Europe through Moorish Spain. The earliest definite reference to guns comes from Florence in Italy in the 14th century and the earliest illustration of a gun is to be found in a 14th-century English royal manuscript. These early guns have characteristics that can be traced throughout the history of firearms.

ABOVE Roger Bacon, English experimental scientist, philosopher and Franciscan friar. His 13th-century manuscript included an anagram which contained the formula for gunpowder. Scholars such as Bacon received much knowledge from the East via Moorish Spain.

The early history of gunpowder is as confusing as it is interesting. India, China, the Arabs and Western Europe have all laid claim to its invention. Gunpowder was more widely known to consist of saltpetre, charcoal and sulphur. Sifting through the evidence reveals that, while gunpowder was known to Chinese alchemists as early as the 11th century, they were looking for a material that burned rather than propelled missiles. While alchemists of different cultures may have known about gunpowder they may not have understood the correct proportions to use to create explosions.

Early evidence

The earliest hard evidence for gunpowder is found in European manuscripts of the 13th century. There are references to gunpowder in the *Liber Ignium* of Marcus Graecus, which dates from about 1300, and in the treatise *De Mirabilium Mundi* of Albertus Magnus who died in Cologne in 1280. The best known is in a manuscript of Roger Bacon, written in about 1260. In this *Epistolae de secretis operibus artis et naturae et de nullitate magial* is a Latin anagram that, when solved, can be translated as "seven parts of saltpetre, five of new hazelwood and five of charcoal". This is an effective recipe for gunpowder. There are also references in the *Opus Majus* of 1267, and in the *Opus Tertium* of *c.*1266–68 which describes a form of firework. The method calls for the powder to be enclosed in an instrument of parchment the size of the little finger, somewhat reminiscent of later paper cartridges.

If gunpowder originated in China, it probably came to Western Europe through the alchemical works of Moorish Spain during the first half of the 13th century. The role of the Arabs in the early history of gunpowder and firearms has probably been underestimated. An Arabic treatise found in Leningrad describes arrows or bullets being fired through a tube by gunpowder, but the date of the manuscript has been disputed, so the claim that firearms were invented by the Arabs cannot, at present, be substantiated.

The Milemete gun

Some early references to guns related to military campaigns. In 14th- and 15th-century Europe, the Hundred Years' War was the most extended conflict, lasting until 1453.

In one of his early incursions into Scotland in his campaign to take the Scottish crown, Edward III of England may have had with him an early form of gun. In a manuscript dated 1326, its author, scholar Walter de Milemete, included a depiction of a vase-shaped gun supported by a trestle. This type of gun was to become more and more commonplace. Protruding from the vase was a large metal arrow. Some kind of wadding was probably placed around its shaft in order to maximize the propellent force. The vase itself would have been formed from cast iron or bronze. The weapon was fired by applying a red-hot iron or something similar to a touch-hole to ignite it. Such devices would come into ever more frequent use from this time onwards.

RIGHT The earliest-dated picture of a gun, from the manuscript *De Notabilitatibus, Sapientis et Prudentia* of Walter de Milemete, 1326. More efficient systems of firing guns were soon devised.

The first gun and prototypes

According to legend, the gun was invented by a German monk known as "Black Berthold", who came from Freiburg in southern Germany. While he was making gunpowder with a mortar and pestle, the mixture exploded, propelling the pestle from the mortar like a bullet, thus giving him the idea for a gun. The monk is a purely legendary figure, but this story of the introduction of firearms may contain an element of truth as it is recorded as early as the 15th century.

There are early accounts of prototype guns in both Chinese and Arab sources. From a 12th-century Chinese source is a description of a bamboo tube filled with gunpowder, which fired arrows. It was mentioned in a military treatise of the 17th century but is based on much earlier sources. Here it was described as a long copper tube containing an arrow, the stock being a short pole. A very similar primitive gun is shown in a 14th-century Arabic manuscript. Many of the early guns were designed to fire arrows rather than other missiles.

In 1326 the Council of Florence decreed that two men should be appointed to manufacture "iron bullets or arrows and metal cannon". Further references of

the same year, also from Florence, mention "cannon, iron balls and gunpowder". The earliest precisely datable illustration of a gun, the Milemete gun, also dates from 1326.

The very marked flask-shape of the gun is not just artistic licence but is based on the sound principal of reinforcing the gun at the back, where the initial explosion happens. This lesson was learned the hard way, when guns had burst at this weak point.

RIGHT Berthold Schwartz was a fictitious German monk known as "Black Berthold". He was said to have discovered gunpowder around 1320 after experimentation with solidifying quicksilver.

Early cannons and handguns

It was soon discovered that the vase-like shape of the early guns, like the Milemete gun, was not the most efficient, and that a traditional projectile like an arrow was not best suited to the new invention. A straight, metal tube was soon developed. The next step was to find the most efficient method of lighting the gunpowder. A method of absorbing "recoil" – the force of the gun going backward after the shot was fired – also needed to be found. The effectiveness of these weapons was growing – the handgun, as we know it, had arrived.

First developments

The Milemete gun demonstrated a stage in development whereby the notion of an arrow as a projectile is still in mind. The shape of the arrow, however, was not suited to maximizing the propellant effect of the explosion and, by the end of the 14th century, stone cannon balls were almost always used.

The Loshult gun

The vase-like feature occurs on the earliest known hand cannon, the Loshult gun, excavated in Sweden in 1861. Cast in bronze, it has a distinctive flask-like shape and also a simple counter-sunk pan around the touch hole to help ignition. It has been dated to the first half of the 14th century. A gun found in the sea near Morko, Sweden, has also been dated to the 14th century. Cast in bronze with a two-stage polygonal barrel, there is a cast bearded figure behind the square pan or touch-hole and a projecting hook below the breech. The hook stopped the recoil when the gun was

ABOVE An early iron gun. This long, thin hand cannon is from Vedelspang Castle, Copenhagen. Dated around 1400, it was more dangerous if used as a club.

fired, hence the name "hakenbüchs" (hook can). The surface was engraved with religious texts. The gun was very short and was clearly made as a handgun.

Another early handgun was excavated from Tannenberg Castle in Germany, destroyed in 1399. Made of cast bronze, the gun had a two-stage polygonal barrel. The bore, in front of the powder-chamber where the missile is held, has deliberately been made narrow, probably to increase the force of the explosion. The Tannenberg gun dates from about 1350.

BELOW An early gun, dated around 1350, was found near Loshult, Sweden. The Loshult gun looked like a miniature cannon and would have been fired in much the same way as the gun depicted in the Milemete manuscript, i.e. from a trestle.

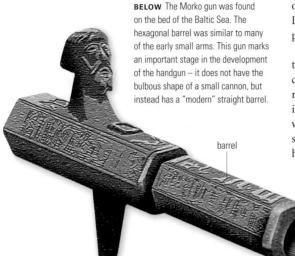

BELOW The Morko gun was found on the bed of the Baltic Sea. The hexagonal barrel was similar to many of the early small arms. This gun marks an important stage in the development of the handgun – it does not have the bulbous shape of a small cannon, but instead has a "modern" straight barrel.

barrel

or twine soaked in a potassium nitrate solution. It was then left to dry. When lit, it would burn at a predictably slow rate.

It is thought that a bullet fired from a gun of this type was capable of penetrating the steel armour commonly used at the time, within a reasonable range. This was in addition to the psychological impact on the enemy of the noise and flash of the weapon. The effectiveness of these comparatively simple weapons was not underestimated and the handgun was here to stay.

Continuing development

As the understanding of the forces involved grew, so did the power and size of the weapons. In due course new designs began to appear to deal with the force of the recoil. Sometimes the handgun was supported on a tripod. Occasionally the stock was not made of wood but was instead formed of a curled metal extension to the main barrel.

Design and construction

Both the Milemete and Loshult guns looked more like cannons than handguns and they also comprised a bulbous area at the back. The Morko gun, dating from about the middle of the 14th century, was more recognizable as a handgun, with a straight firing tube, and it was designed to be fired without the use of any supports, such as a trestle. The Morko gun, like the Tannenberg gun, was a straightforward metal tube attached to a wooden stock, which was either a pole or a construction similar to that used for a crossbow. Weapons such as these may have been designed to accept a metal bolt but the Tannenberg gun when excavated was found to be loaded with a lead bullet. A variety of ancient handguns has been found in various locations, some of them in Spain, others in Switzerland, Sweden and England.

Firing the early guns

The metal tube (or barrel) of a handgun such as the Morko gun or the Tannenberg gun was about 300mm/12in in length. The bore (the interior diameter of the barrel) was about 17mm/0.67in. A handgun had a powder chamber, which was linked to a small hole drilled at right angles. The gunpowder and propellant would have been inserted through the muzzle (the forward end). The gun was probably fired by inserting a red hot wire poker into the touch-hole. Another way of firing would have been to use a slow-burning match fuse. The slow match was usually made up of cord

ABOVE Use of handguns in the 15th century as depicted in *Rudimentum Noviciorum*, Lubeck, 1475. This gun was mounted on a pole and appeared to be lit through a touch-hole.

Matchlocks

The matchlock was the first major step forward in developing an efficient firing mechanism that would allow the user to concentrate on holding his weapon with both hands and aiming it effectively. As the power and efficiency of the handgun increased, the wooden stock began to be developed, so that the recoil could be effectively absorbed. Increased weight meant that a pike or forked support between the weapon and ground were used.

The serpentine lock

The method of manually igniting the charge by means of a rod that held a piece of metal or glowing tinder, was superseded in the early 15th century by a device known as a serpentine (an angled arm) or serpentine lock. The first mechanical firing mechanism was the matchlock, the basis of which was the serpentine. The mechanism had to be both efficient and robust.

The early matchlock was operated by pressing the lever up to the stock so the match or tinder was pushed into the touch-hole. Since the pole stock could be placed under the arm at the shoulder, the gun could now be aimed properly. This development led to changes in the shape of the stock, which became polygonal. Some handguns, however, retained the pole-stock until well into the 15th century. It is possible that the serpentine lock was suggested by the

BELOW A drawing of parts of a matchlock. The simple mechanism could work well as long as match and powder were dry.

trigger action of a crossbow, which, like the serpentine, was a long lever under the stock which was pressed upwards to release the cord. The polygonal profile of some early gun stocks also resembles those of crossbows. The crossbow was the most powerful hand-held missile weapon then in use, so it was logical for early gunmakers to copy its features.

Serpentine development

The first recorded illustration of a serpentine lock is from a manuscript dated 1411 (Codex Vindobana 3069) which can be found in the Austrian National Library. This shows the simplicity of the serpentine system, and the straightforward operation of the lever mechanism. The man firing the weapon is able to keep both hands on the stock making aiming and firing easier. It is known that stocks grew in length in order to absorb recoil from firing, and they ranged between just under a metre to 1.5 metres (1.64 yards) in length. As stocks developed, different methods of holding the weapon also evolved.

LEFT The gun-wielder from this 14th-century manuscript is firing the weapon from under his arm, but over time firing from the shoulder became more common. The kneeling figure is in the process of casting bullets.

Some matchlock serpentines included a spring, which would move the arm holding the match more rapidly into the pan. The "match" was normally a length of woven hemp cord boiled in a solution of saltpetre and then allowed to dry. It was literally looped around the fingers. It was carried lit, smouldering ready to ignite the gunpowder. The invention of the matchlock allowed mechanical means to press the glowing end into the touch-hole. In addition the matchlock was a simple mechanism, easy to make and repair and reasonably reliable. It did not, however, resolve the problem of how to ignite the powder without having to carry a constantly glowing match. It also could give away the position of the user to the enemy.

Fifteenth-century improvements

During the 15th century a number of improvements were made. The touch-hole was moved from the top of the barrel to the side. This led to an improved pan, which was mounted on the side of the barrel and had a hinged cover to protect the powder. As the arm holding the match descended, the flash pan cover moved aside exposing the powder in the pan. Once the powder had been ignited and the gun fired, the pan cover would return to its original position. This ingenious system gave the powder in the pan much greater protection from the effects of wind or rain or from being inadvertently tipped out of the pan. It also meant that the weapon was safeguarded against sparks inadvertently firing the gun before the

firer was ready. More rapid action was achieved by the development of the "snap-matchlock" in which the serpentine arm was replaced by a curved match-holder driven by a spring.

BELOW An ornate matchlock mechanism from the 18th century. Despite the ornate design, the matchlock was operated by the simple process of a lit match igniting the gunpowder.

ABOVE A handgun being discharged. This is a matchlock handgun with a pike rest. Some matchlocks had a disproportionately large stock requiring a support to take the weight.

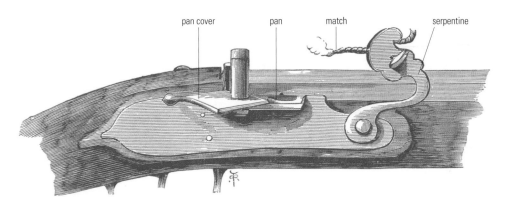

pan cover   pan   match   serpentine

RIGHT Cross-section through a firearm showing the internal workings of the snap-matchlock: when the trigger lever was pressed the lighted match fell into the powder in the pan and ignited the charge. The curved matchholder was drawn back against a spring so that when the trigger was pulled the match came down very quickly to the powder.

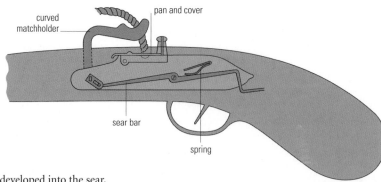

curved matchholder

pan and cover

sear bar

spring

The snap-matchlock

By about 1470 the system had developed into the sear, or snap-matchlock. The serpentine was replaced by a curved matchholder driven by a spring. When the trigger lever was pressed, the lighted match fell into the powder in the pan and ignited the charge. The match was held down in the pan through the action of springs and levers mounted on the inside of the lock-plate. The matcholder was then pulled back manually and was held by the sear (a lever or catch). When the trigger was pulled, the sear was withdrawn and the matcholder snapped back into the pan. The mechanism was soon superseded by new inventions in Europe, but was taken to Japan by Portuguese traders in the 16th century and used in Japan until the 19th century. The Japanese snap-matchlock pistol is a scaled-down version of a long arm, fitted with a wide barrel and the matchlock on the side of the stock.

Advances in gunpowder

The gunpowder used at the time was also known as serpentine. One defect of serpentine powder was that its different constituents, of different mass, tended to separate over time or during movement. The saltpetre and sulphur would settle at the bottom and the charcoal element at the top. This was likely to make ignition less

efficient. It was only with the invention of "corned" powder in the early 15th century that this problem was addressed. In this process, the ingredients were all ground and mixed together and then sieved to make sure that all the constituent particles were the same size.

The matchlock in warfare

By the 15th century firearms were used to an increasing extent on the battlefield. Improved systems of ignition such as the serpentine lock and the snap-matchlock, together with changes in the shape of guns, made them lighter and easier to use. Improved methods of making barrels included the introduction of rifling within the barrel of the firearm. Rifling refers to the grooves that were created in the barrel of a firearm to produce a spin on the projectile to improve accuracy of the shot.

By the end of the 15th century firearms were playing an increasing role in warfare. The matchlock musket gave the common footsoldier a much better chance of bringing down a swordsman without either of them having to be in contact. The weapon stock began to change to a flattened form that could be placed against the shoulder, one hand gripping the butt and operating the trigger while the other supported the barrel. Assuming the powder did not simply "flash in the pan" (the explosion of gunpowder in the pan), but ignited the charge, in the breech, the bullet was propelled in a cloud of smoke and with a loud bang.

However, the disadvantage was that the matchlock musket was much better suited to use by the infantry than the cavalry. The cavalry needed pistols that could be fired on horseback and could be lit without using a

LEFT An early 17th-century musketeer fires a matchlock musket. The musket replaced the arquebus, a firearm used from the 15th to the 17th centuries. The matchlock gradually became lighter and more compact in design and more efficient to use.

ABOVE Matchlock musket from around the late 17th century as painted by the circle of the Master of Calamarca.

constantly glowing match that was exposed to the weather. However, matchlocks were cheap and simple to produce, and so stayed as part of armed warfare on the battlefield for over 200 years. Although considered obsolete in the 16th century, some matchlocks were still being used during the 18th century.

Fire-lock mechanisms

Throughout history gunmakers have attempted to improve their weapons and make them easier to hold and shoot. After the development of the matchlock there was a move towards producing something more elegant with a more effective method of ignition. What was required was a device to produce a fire on demand. In the 16th century matchlocks were replaced by fire-lock mechanisms (so-named because they created fire by striking pyrite against steel). Early versions used the wheel-lock mechanism, but the more efficient flintlock mechanism soon followed.

Development of the pistol

Although some of the early matchlocks were small, essentially they were not "small arms". They were more closely aligned with small shoulders – or long arms. Early pistols were often referred to as "small arquebusses". The arquebus was a small firearm used during the 15th to 17th centuries, using the matchlock mechanism. The arquebus would later slowly be succeeded by the larger musket, a muzzle-loading shoulder gun with a long barrel, beginning in the 16th century. Surviving examples of small handguns or pistols fitted with matchlocks are extremely rare and were not widely used, except in Japan where the snap-matchlock was in common use. The term "pistol" did not come into general use before the 16th century. Its origins are unclear. It may have developed from the word *pistolet* which in the 14th century meant both a dagger and a firearm.

ABOVE A Japanese matchlock musket from the 18th century. The snap-matchlock mechanism was used in Japan until the 19th century.

The wheel-lock

In the early years of the 16th century there came an invention that was a milestone in the history of firearms – the wheel-lock. The matchlock had relied on a lighted match which was awkward to hold and entirely susceptible to the weather. It was normally held between the fingers and was dangerous to use. Seventeenth-century accounts tell of careless gunners going to replenish powder, forgetting that they were holding a match, and as a result blowing up entire powder magazines. The wheel-lock solved this problem, as the gun could simply be aimed and fired. It remained in use until about 1650.

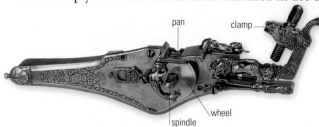

pan   clamp   wheel   spindle

LEFT An example of the revolutionary wheel-lock mechanism. Invented in the early 16th century, this development represented a firearm milestone. It was safer and more convenient than a matchlock, since there was no longer a need to keep a smouldering match close to loose gunpowder.

How a wheel-lock worked

The wheel-lock worked like a giant cigarette lighter. A wheel cut with slots turned against a piece of iron pyrite, producing a shower of sparks that ignited the powder in the pan. The wheel was turned by a short chain. Protruding from the wheel was a spindle. A spanner was used to wind the mechanism and this depressed the spring and wound the chain round the spindle. When the wheel was fully wound it was locked by a series of levers. Pulling the trigger withdrew the locking lever and the wheel then revolved around the pyrite. The pyrite was held in jaws at the end of a curved arm, which was lowered manually against the wheel. The pan had a sliding cover operated by a button. In later wheel-locks the pan-cover slid back automatically when the trigger was pulled.

An Italian invention

Recent research by arms historian Claude Blair has established that the wheel-lock was almost certainly invented and first used in the province of Friuli in north-east Italy, in the early years of the 16th century. The letters of Luigi da Porto (1485–1529), an author and military commander in Venice, provided the evidence. In 1510 he was on campaign in Friuli and visited the city of Cividale where he observed some of the citizens using handguns for "shooting little birds high in flight" and also fish in clear water. He also describes an action between Venetian and Imperial German forces in which "little guns", which resembled iron maces, were used to great effect. He describes the guns as being "three spars long", although this does not translate to any known measurement.

RIGHT Cross-section through a firearm showing the workings of the wheel-lock. A spring was connected to the wheel's axle by a short length of chain. The axle was wound back to put the spring under tension. When ready to fire, the trigger was pulled. The wheel was released and would spin rapidly, scraping against the pyrite. White-hot particles from the pyrite fell into the flash pan, igniting its powder and then the main charge.

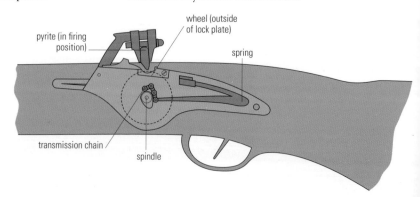

pyrite (in firing position)   wheel (outside of lock plate)   spring   transmission chain   spindle

BELOW French wheel-lock
pistol. The elaborate engraving
and high-quality materials
are typical of French pistol
manufacture. They reflect the
complexity of the wheel-lock
mechanism, which required a
high level of craftsmanship.

RIGHT Wheel-lock six-shot
revolver from Germany, dated
around 1600. This multi-barrelled
firearm has a six-chambered
cylinder that needs to be rotated
by hand.

It is almost impossible to shoot birds in flight or
fish in water with a matchlock, as the "flash in the
pan" (the explosion of gunpowder in the pan) acted as
a warning. This led inevitably to the conclusion that
the little guns must have been wheel-locks. As Friuli
province belonged to Venice, it also explains why some
of the earliest examples of the wheel-lock are to be
found in the Doge's armoury in the Palazzo Ducale,
Venice. Two of these are horseman's axes, perhaps the
"iron maces" that Luigi da Porto saw in Cividale. These
have been dated to about 1520.

The wheel-lock in Germany

German gunsmiths also seem to have adopted the
wheel-lock early in the 16th century, as there is an
account in 1507 of a servant being given money by the
steward of Cardinal Ippolite d'Est to go to Germany
and buy "a gun of the type that is ignited with a stone".
Another reference dating from 1515 describes a young
German accidentally shooting a girl in the neck with a
gun that "fired itself". An indication of how novel and
dangerous wheel-locks were considered to be in the
early 16th century is the edict issued in 1517 by
Maximilian I (1459–1519) banning their use in the
Habsburg empire.

The earliest-dated wheel-lock pistol was made for
the Emperor Charles V (1519–56), almost certainly

by a German gunsmith. It is dated 1534 and has a
stock shaped like the grip of a dagger. It still has its
original key. Although the mechanism remained
basically the same, there were changes to the shape of
the stock of wheel-lock pistols. By the end of the 16th
century the stock was angular; one type of German
wheel-lock, the "puffer", which was fashionable in the
1580s, had a stock terminating in a large ball. This
allowed it to be drawn from large holsters, which were
then fashionable. Wheel-locks were in use until the last
quarter of the 17th century, but were being superseded
by other forms of ignition by the 1650s.

RIGHT A horse soldier with a wheel-lock pistol, from the *Kunstbuchlin*
by Jost Amman, Frankfurt, 1599. Although the development of the
wheel-lock was not widely adopted because they were expensive to
produce, they allowed pistols to be aimed and fired with one hand.

The snaphance

An important development in the history of firearms, the snaphance led directly to the invention of the flintlock. The snaphance was an improvement on most of the previous systems of ignition. It was almost certainly a German invention based on the snap-matchlock. Invented in the 1540s, it was rapidly adopted throughout Europe. Local variants of the snaphance lock were developed throughout the 17th century.

Early snaphance designs

In the 16th century a suitable replacement for the wheel-lock needed to be found for military use. It required something cheaper and simpler than the wheel-lock. The snaphance lock was invented in about 1540. A Florentine ordnance dated 1547 describes some guns having locks of this form. The earliest snaphance lock known is to be found on a gun in the Royal Armoury, Stockholm. This is one of a group of thirty-five documented as having been fitted with these new locks at the Royal workshops at Arboga in central Sweden in 1556. The miquelet was a Mediterranean version of the snaphance lock. Both the snaphance lock and the miquelet played an important part in the development of the flintlock, the next step

BELOW Part of a snaphance lock from about 1580. The action of the mechanism was likened to a cock or hen pecking at grain, possibly the origin of the word "cock" for the arm of the mechanism.

clamp

cock

steel plate

forward in the development of firearms.

The name of the snaphance lock probably derived from a Dutch expression describing its action – like that of a pecking hen or cock ("snap haan"). Although this suggested a Dutch origin for the lock, the evidence seems to contradict this, as some of the earliest known snaphances are German and this type of lock was almost certainly a German improvement on the earlier snap-matchlock.

The workings of the snaphance

In the snaphance lock, the cock was fitted with jaws that hold flint or iron pyrite. The cock was pulled back against a large spring and when the trigger was pulled the cock fell and the pyrites or flint struck a pivoted arm known as the steel plate (or battery). This produced sparks which ignited the powder in the pan. In its earliest form the pan-cover had to be opened before pressing the trigger. There were variations of the lock throughout Europe.

BELOW Cross-section through a firearm showing the workings of the snaphance lock. The lock was invented in around 1540. In the snaphance system, the flint was held in a clamp at the end of the cock. Once the trigger was pulled, the cock was released and, under strong pressure, the flint struck against the steel plate (or battery). White-hot steel shavings were produced that fell into the flash pan and ignited the priming powder. The steel could be retained in a "safe" position, forward of the pan if required. The cock could not then cause a spark if accidentally released.

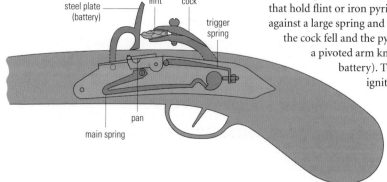

steel plate (battery)

flint

cock

trigger spring

pan

main spring

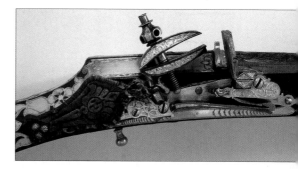

The miquelet or patilla

Another form of snaphance was the Mediterranean lock or miquelet, widely used in southern Europe and northern Africa. The Spanish version of this lock was known as a patilla lock. At least one writer on Spanish firearms in the late 18th century declared that the miquelet was a Spanish invention, although the evidence contradicts this. The snaphance was probably introduced to Spain in the 1570s. The main characteristics of the Spanish snaphance were the large external mainspring, which acted on the heel of the cock, and the steel plate (battery) made in two parts so that the front could be replaced. The steel plate was cut with a series of vertical grooves. The most noticeable feature, however, is the form of the jaws, which were long and rectangular and held together by a screw with a large open ring at the top.

The snaphance was most widely used in Italy. It had its own local variation – the Roman lock – which had a mainspring operating on the end of the cock, and two sears to allow for the half-cock and full-cock position. It had a horizontal sear and the usual separate pan-cover and steel plate. This type of lock possibly derived from English locks, which were very popular in Italy in the 17th century.

ABOVE A miquelet lock sporting gun from Algeria. Sometimes known as the Mediterranean lock, the miquelet was characterized by large jaws that held the flint, and by a large main spring.

The Baltic lock

Snapping-lock guns were mechanically simpler and much more reliable than wheel-locks. One of the earliest forms of snaphance was known as the Baltic lock, characterized by prominent jaws held by a screw, a cock formed as a long bar with shallow downward curve and an extension at the back. This extension caught on a sear which operated horizontally through the lock. The mainspring was mounted on the outside of the lock-plate.

The Netherlandish lock

There were local variations of the snaphance produced in the Low Countries, and in England and Scotland. The Netherlandish lock had a cock with a projection at the rear, which connected with a sear. The pan-cover opened automatically when the lock fell and the steel was attached to the end of a pivoting arm. The Netherlands developed a trade in the lock.

The Scottish lock

There is some evidence to show that the snaphance was known in the early 17th century as a Scottish lock. With their engraved brass stocks and locks, Scottish firearms have always been appreciated as curiosities and enjoyed wide distribution in the 16th and 17th centuries. The two earliest Scottish snaphances known are a pair with left- and right-handed locks in the armoury of the Electors of Saxony in Dresden. They are dated 1598.

The English lock

A type of snaphance found on English firearms of the 1630s is known as the English lock. This could be engaged at the half-cock position. This was a safety device for the lock. Like the Baltic lock, the English lock has a back-catch, which is caught in a notch cut into the back of the cock.

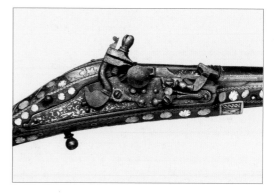

ABOVE A highly decorated 16th century snaphance. The pistol shows the mechanism's high level of sophistication. The flintlock that was developed during the 17th century was less complex to produce.

The early flintlock

The invention of the flintlock in the early 17th century brought together the best features of the local snaphance designs. The flintlock was comparatively simple and very efficient, and was widely used on pistols and long guns. It originated in France but, by the middle of the 17th century, it was in use in most European countries. Because of its popularity the flintlock was in service for over 200 years and was still being made in the 1830s.

The first flintlocks

The flintlock was a development of the earlier snaphance. The early mechanism, sometimes called a doglock, has the steel and pan-cover made as one unit. The sear moves vertically, engaging in two notches cut in the tumbler (a steel cam attached to the axis shaft of the cock). These notches give either half-cock or full-cock as required, the safety position being half-cock.

ABOVE An English flintlock mechanism from London of around 1775. The flintlock mechanism was relatively straightforward and reliable. It became the primary means of ignition for muskets and pistols for the best part of 300 years.

To fire the weapon the cock was pulled back to full-cock and the trigger was pulled. The flint struck the steel, which opened the pan lid and created a spark which fell into the powder. The powder ignited and set off the main charge of powder which was placed in the barrel, with the propellant, such as a ball.

The flintlock was almost certainly invented by a member of the le Bourgeoys family who worked in Lisieux, Normandy. Jean, Pierre and Marin le Bourgeoys had a workshop which specialized in the production of mechanical devices such as locks and watches, and Marin was mechanic to the king. The earliest datable flintlock mechanism is to be found on a French royal gun from the famous cabinet d'armes of Louis XIII. The stock was inlaid with the royal cipher and the barrel was stamped with the mark of Pierre le Bourgeoys. As Pierre died in 1627, the flintlock is likely to have originated sometime in the 1620s.

The earliest recorded pistol using the flintlock system is a revolver in the Kremlin armoury, Moscow. It was made by a Russian gunmaker named Pervusha Isaev who worked in the Kremlin armoury around 1625 and had clearly come across contemporary French flintlocks.

RIGHT Cross-section through a firearm showing the workings of the flintlock. The flash pan, or frizzen, is filled with priming powder and then closed to protect it. The cock holding the flint is then drawn back to half-cock. In this mode the weapon cannot be fired. To fire the weapon the cock is pulled back to full-cock. When the trigger is pulled the flint strikes the steel, which opens the pan lid and creates a spark which falls into the powder. The powder ignites and sets off the main charge, which has been placed in the barrel.

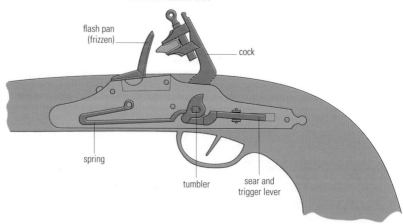

flash pan (frizzen)

cock

spring

tumbler

sear and trigger lever

Common features

Flintlocks dating from the first half of the 17th century have several easily recognizable characteristics. The lock-plate was flat, attached to the stock by two screws, and the upper and lower edges were usually parallel. In early examples the lock-plate extended downwards at the centre, revealing a feature found on the earlier wheel-locks. The cock was flat, with a long, curved narrow neck and forward jaws to hold the flint. Early flintlock pistols were very long, the barrels supported by a full stock which was flat and curved gently at the butt-end, which made the early pistols appear very graceful. Initially the butt-cap (its base) had a short spur extending down the stock. By the end of the 17th century this spur extended almost to the lock-plate. In about 1630 a lock-plate and cock of concave section were introduced, for added strength. After 1660 the profile of the lock-plate became concave. It is these features that enable the specialist to date flintlock pistols accurately.

National characteristics

With the spread of the flintlock across Western Europe after the 1630s, national characteristics began to emerge. Each workshop had its own special, easily recognizable style. France dominated the firearms trade in the early 17th century and many fine flintlock pistols that are preserved in the famous armouries of Europe were made by the gunmakers of Paris.

In Scotland the true flintlock was not introduced until about 1700, the earlier snaphance being preferred. The shape of Scottish pistols was always very distinct and the flintlock is found on pistols with the characteristic heart-shape and ram's horn butt.

By the middle of the 17th century the form of the flintlock was fully developed and it changed little over the next century. One technical feature that was found on some later flintlocks was the *detent*. This was a small, hinged lever attached to a tumbler, the mechanism which held or released the power of the mainspring on the cock. The *detent* stopped the sear from going to the half-cock position as the cock fell.

Decoration

The stock of a flintlock pistol could be elaborately carved and decorated. One very distinctive group of pistols made by the gunmakers of Maastricht in the Low Countries in the 1650s have stocks of carved ivory, the butt-caps terminating in carved classical heads with helmets in the Baroque style.

Italian flintlock pistols of the latter part of the 17th century are also very distinctive. Gunmaking centres like Brescia in northern Italy were famous for their firearms. A feature of their work is the use of chiselled, engraved and pierced steel to decorate the stock. The locks are chiselled in low relief with flowers and animals, and the butt-plates are decorated with finely pierced and engraved mounts. Engraving was a decorative technique widely used on the mounts and locks of firearms because it did not weaken the working parts by removing too much metal from the structure.

Compared to the work of continental gunsmiths, the products of English makers during the 17th century were comparatively modest, the decoration being usually limited to rather naïve engraving, although pierced silver butt-caps are occasionally found on late 17th-century English flintlocks.

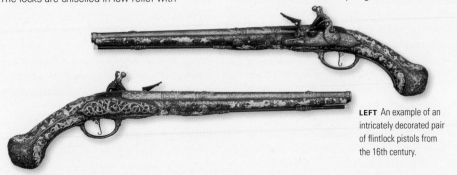

LEFT An example of an intricately decorated pair of flintlock pistols from the 16th century.

Breech-loaders and early revolvers

As the development of the handgun continued, the minds of both craftsmen and gun-users turned to the problem of how to fire more shots while taking the least amount of time to reload. Both the breech-loading system for pistols and early revolvers were attempts to improve efficiency in this area.

BELOW A flintlock breech-loading magazine pistol from 1780. The use of breech-loading mechanisms demonstrated the continual quest to increase the speed of reloading; however, with powder and ball they were rarely successful.

Breech-loading

There are early documentary references to breech-loading guns by the middle of the 14th century. A late 15th-century manuscript from the University Library at Evlangen in Germany illustrates some breech-loading handguns with separate reloadable chambers. The earliest surviving breech-loading firearms date from the early 16th century. These generally have a hinged section at the breech which can be opened to allow a reloadable chamber containing powder and ball to be inserted. A carbine (a small firearm) made for Henry VIII of England, dated 1537, is of this type.

The turn-off barrel

Another form of breech-loader used a system in which the barrel could be unscrewed. This was widely used in England during the 17th and 18th centuries, particularly on "Queen Anne"-style flintlock pistols. It became apparent that greater accuracy could be achieved if a large ball was used with a rifled barrel that unscrewed. A rifled barrel had a number of grooves that have been cut, pressed or forged into the barrel of the weapon to stabilize the projectile. It is recorded that Prince Rupert hit the weathercock of St Mary's church in Stafford, England in 1642, twice in succession, using rifle-barrelled pistols.

Barrels could also be hinged so that they "broke" rather like a modern shotgun. When the trigger-guard was pulled back it revealed a locking-catch on the top of the barrel. The charge and ball were located in a separate reloadable chamber. The best breech-loading systems relied on a vertical plug which could be unscrewed by turning the trigger-guard. Patents were taken out in the 18th century for this system but they were used almost entirely on long guns rather than pistols. Patents were also taken out in the 19th century for barrels that swivelled and chambers that pivoted.

The Roman-candle system

The problem of making a firearm that could fire a series of shots was solved initially by increasing the number of barrels. However, by the 16th century the idea of the super-imposed load was developed. It was known as the "Roman-candle system", named after a firework that threw out a series of burning balls. A series of charges were placed next to each other, the bullets being pierced and loaded with fuse. In theory, when the first charge was fired, the ignition should have passed to the one behind it, firing the charges in a measured series. This system was used on specially designed wheel-lock pistols which fired three shots, but needed three separate locks.

French four-shot repeater

In the 1630s a clockmaker of Grenoble in France produced a repeater that could fire four shots from a single barrel. In order to prevent all the charges going off together – a major problem with superimposed loads – the pistol had to be carefully loaded with tight-fitting wads and balls to separate each charge.

The sliding lock
Another system developed in the 1780s for flintlock pistols was known as the sliding lock. Four shots could be fired from four touch-holes, the lock being slid along the lock plate on a bar to connect with each touch-hole.

Magazine systems
Before the invention of the cartridge that contained powder, ball and primer, a repeater had to have separate "magazines" for the powder and ball. In the mid-17th century a Florentine gunmaker, Michele Lorenzoni, developed a repeating system for pistols that used separate magazines for the powder and ball. It also had a priming magazine on the lock. The pistol was primed and cocked by means of a lever on the side, powder and ball passing into the chamber when the pistol was raised and lowered. Versions of this system were made in England as late as 1800.

The magazine system developed by the Kalthoff family in the 1640s is only found on long guns, although a version was used on pistols made by the German gunsmith Sigmund Klett.

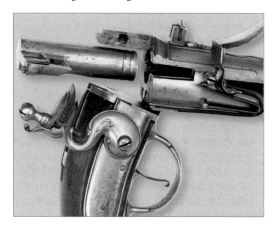

ABOVE A flintlock breech-loading gun from around 1690. Breech-loading of the powder and ball led the way to new developments such as the metal cartridge, shown above, which would replace the powder, ball and primer and eventually make the flintlock mechanism obsolete.

The early revolver

LEFT The seven-barrel revolving flintlock. The introduction of revolving barrels meant it was no longer necessary to reload after each shot, though the primer had to be replaced in the flash pan. The system would work better with the percussion cap lock.

The earliest example of a revolver is generally accepted to be a three-barrelled matchlock pistol in the Palazzo Ducale, Venice, which dates from about 1540. The barrels are turned manually and locked in place by a spring catch. Another early revolver is the three-barrelled wheel-lock pistol etched with devices associated with the Emperor Charles V. The pistol is of steel and fires steel darts from small-calibre barrels. A wing-nut at the end of the stock allows the barrels to be moved and locked. Firearms with revolving cylinders and single barrels were known by 1600, and the 17th and 18th centuries saw several ingenious versions. One of the best known is the six-chambered snaphance revolver made by John Dafte in the 1680s. It has a brass barrel and cylinder and six chambers, each of which aligns with the barrel when the pistol is cocked. The Swiss gunsmith James Gorgo also made a revolver in the late 17th century with three chambers, the cylinder being released by pulling up the trigger guard. The late 18th century saw the development of the pepperbox (an early form of revolving repeating pistol), but it had been anticipated in the 1730s with the five-barrel pistol made by Johan Gottfried Kolbe. It was manually cocked for each shot, the barrels rotating when the pistol was cocked.

Flintlock development

The invention of the flintlock was to have a transforming effect on world history. Although imperfect in many ways and liable to misfire, the flintlock's success was based on its essential simplicity and ease of maintenance. The testament to its success was that the flintlock remained the basic ignition system for a pistol or musket for over 200 years.

RIGHT A Royal Navy Sea Service pistol from 1805. A typical pistol of this kind would have a walnut stock and be fitted with a brass trigger guard and a brass butt cap. The elegant simplicity of the design also speaks of reliability, essential for seamen in battle.

Flintlock pistols

The flintlock pistol had a comparatively short range and, as it was often not rifled, it was also not very accurate. It was normally used as a close-range back-up to the sword or cutlass, as a self-defence weapon, or for duelling.

During the 18th century, a pattern of Sea Service pistols began to emerge in the Royal Navy, and these were often used by boarding parties, in addition to the cutlass, or they could be used to repel boarders. These pistols, some of which were manufactured by Richard Wilson in England, were either produced in 300mm/12in long form or 225mm/9in short form, the latter being better suited to the mêlée of a boarding

ABOVE The flintlock pistol was highly regarded by pirates and carried when boarding ships. However, reloading was so slow that pirates often used the pistol's butt as a club.

party. A typical Sea Service pistol would have a walnut stock, fully stocked to the muzzle, and would be fitted with both a brass trigger guard and a brass butt cap. It would have a pipe for the ramrod (which was inserted into the gun's barrel). The ramrod itself was made of wood and had a brass tip. The pistol would have a steel trigger and there was also a steel belt hook fitted on the side. The hook would have made the pistol easy to carry and to replace on the belt when it had been fired.

Apart from its use by the navy, the flintlock pistol was also used by cavalry in the 18th and 19th centuries. It was, of course, impossible to reload while in the midst of a charge or a mêlée with the enemy, so the pistol would normally be fired in the first instance when reasonably close to the enemy lines, to cause as much damage and confusion as possible, and the cavalry cutlass (a short, curving sword) would then be used for close-quarter action.

Like the Sea Service Pistol, the British also introduced a standard type of cavalry pistol in 1796. Called the New Land Cavalry Pistol, it was designed for units such as the Royal Horse Artillery. Fitted with a flat butt and a ring to make it easy to carry when not in use, it also had a ramrod that swivelled to make it easier and quicker to reload when on horseback.

Huguenot gunmakers in England

Compared with guns made in France, which was one of the chief centres of gunmaking in Europe, English guns were relatively plain in design. Strangely enough, this characteristic may not be entirely native to England for many Huguenot gunmakers came to England as refugees in the 17th century and they would have had a significant influence on the design of English guns.

Pocket pistols

Flintlock pistols, however, were not all designed for the derring-do of the battlefield or the high seas. Gunmakers were also conscious of a market that continues to be lively today – for discreet, elegant weapons to be used in self-defence. In the days of highwaymen, this was a useful asset. Bunney of London, for example, is known to have manufactured a 150mm/6in ladies' gold pistol with an elegantly worked stock. Such a weapon could be easily concealed about the person and used to devastating effect if necessary. There is also a man's version in plainer style of the same proportions.

The box lock

Concealment of a flintlock pistol threw up an obvious difficulty. The mechanism was large and unwieldy and would be liable to get snagged, making quick deployment difficult. Gunmakers therefore developed what is known as the "box lock". This was a system whereby the cock and frizzen were removed from the sides and concealed in the body of the pistol, only emerging once the trigger was pulled. The pan was taken away leaving only a saucer-shape depression surrounding the touch-hole. In some versions, the

ABOVE A pair of 18th-century lady's muff pistols made by Charles Gourley of Glasgow. The length of each gun was under 125mm/5in so could be carried easily. They were also louder than the average pistol to frighten off assailants, but would be unlikely to cause serious damage.

trigger and trigger guard were also concealed in the body of the weapon, making it a very easy armament to carry. In this case the trigger was revealed once the weapon was cocked.

Flintlocks in America

The flintlock mechanism in Europe was the mainstay of military forces both on land and at sea for several centuries. It was also hugely important, as a musket or rifle, in colonial America. Although the flintlock mechanism was complex, it was also relatively easy to manufacture, maintain and highly reliable. For the colonists who lived in America in the 17th century it had a huge influence on their ability to survive and to expand westwards. They were used to kill game to feed their families, defend the homestead and to exert dominance over the native American population.

The American flintlock musket evolved from the colonists' weapons. It was effectively used by US forces in the American War of Independence (1775–83) and in the War of 1812, against the English army where it was nicknamed the "Kentucky rifle". The rifle was also the friend of the solitary backwoodsman.

ABOVE *Trade from the Monongahela* by Gayle Hoskins, showing American backwoodsmen buying and testing Kentucky rifles. Whereas blacksmiths had previously been concerned exclusively with horseshoes and similar requirements, the growing importance of firearms led to the development of a skilled cottage industry.

The percussion system

Although the flintlock was a successful mechanism, it also had its limitations. Two factors led to the replacement of the flintlock by a more efficient system. One was the creation of fulminate of mercury and the other was the ingenuity of a Scottish pastor, Alexander John Forsyth, who developed its use in the new percussion system.

Early gunpowder experiments

From the early 17th century, chemists and military specialists had been experimenting with ways to increase the explosive power of gunpowder. Giuliano Bossi, writing in 1625, recommended adding antimony or mercury precipitated with acid. A German alchemist, Johann Kunchel (1630–1703), is thought to have discovered fulminate of mercury, which was later used in the percussion system. There were a number of experiments by French chemists in the 18th century that replaced gunpowder with fulminate of mercury. These were unsuccessful as the fulminates were highly volatile and either detonated too quickly or burst the cannon or firearm in which they were used.

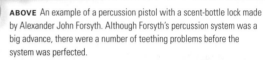

ABOVE An example of a percussion pistol with a scent-bottle lock made by Alexander John Forsyth. Although Forsyth's percussion system was a big advance, there were a number of teething problems before the system was perfected.

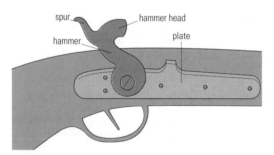

ABOVE Diagram showing the external view of the percussion system. The system was developed to overcome the disadvantages of the flintlock ignition. Many flintlocks were converted to the percussion system. The system is exactly the same as the flintlock in terms of its internal mechanisms and, like the flintlock, the hammer used could be held in uncocked, half-cocked and fully-cocked positions.

RIGHT Wall tablet memorial to Alexander John Forsyth, inventor of the firearm percussion system, in 1807, Tower of London. Forsyth invented the system as a result of a logical necessity to reduce the delay in setting off the main charge.

The work of Forsyth

In 1800 Edward Howard published a paper on mercury fulminate in the *Philosophical Transactions of the Royal Society*. His work came to the notice of an amateur chemist and mechanic of Aberdeen, the Reverend Alexander John Forsyth (1768–1843). Forsyth was also a keen sportsman. Although the flintlock was reliable, it had its faults. The priming powder was easily affected by damp, but more serious from the sporting point of view was the problem of the delay between the priming powder being ignited in the pan and the main charge in the barrel being set off. The puff of smoke from the priming powder frequently alerted the game as well.

Using fulminate

Forsyth realized that if fulminate could be used instead of priming powder, a gun could be fired very quickly once the trigger had been pulled. He had already written a paper on fulminate in 1799 and by 1805 had developed a lock, which used fulminate of mercury instead of priming powder to detonate the main charge. The old flintlock mechanism could be converted to the new percussion system very easily and cheaply. The cock was replaced by a hammer, the steel and spring were removed. The priming pan would be replaced by the circular plug around which the scent-bottle magazine revolved. It was not necessary to alter the interior working of the lock as the hammer, like a cock, was operated by the mainspring.

Forsyth was cousin to the influential characters Henry and James Brougham, who introduced him to Sir Joseph Banks, a patron of science and Member of Parliament. This led to an introduction to Lord Moira, Master General of the Ordnance. Moira realized how useful firearms using the new percussion system would be to the British army, and arranged for Forsyth to have a workshop at the Tower of London. Although initially he received some financial support, Forsyth had difficulty in getting anyone to make the new lock, and also in obtaining sufficient fulminate. A gun-lock and a lock for a small cannon were tried out at Woolwich but the results were disappointing, and an unfavourable report about the new lock was made to the Board of Ordnance. Forsyth was aware that the military potential of his invention could not be immediately realized, especially when a new Master of the Ordnance stopped his work at the Tower.

The 1807 patent

On 11 April 1807 Forsyth took out a patent for the new lock and in 1808, established a business with James Brougham in London at the Forsyth Patent Gun Co. Much of his later career was spent protecting his patent from infringement by competitors.

The scent-bottle lock

To hold the fulminate powder, Forsyth used a container shaped like a flat scent-bottle. It held enough to prime 30 shots. At the top was a striking pin operating against a spring. The container was attached to the lock by a round plug, which allowed it to rotate, and had a channel that led to the touch-hole. The scent-bottle magazine was rotated to drop some fulminate into a cavity on the top of the plug. It was then returned to its original position so that the striker was at the top. Instead of a cock with jaws holding flint, Forsyth devised a hammer that hit the striking pin when the trigger was pulled.

RIGHT The scent-bottle lock was an ingenious device but also potentially dangerous, as the contents could ignite and explode.

The percussion cap

In Britain the percussion lock took some time to establish itself and it was not until about 1820 that it began to replace the flintlock on a large scale. On the Continent, however, the virtues of the new lock were immediately recognized and Parisian gunmakers were soon making copies and taking out patents based on the design of the Forsyth lock. One of the problems of the new percussion lock was how to control the rate of flow of the fulminate, and this led to another significant development – the percussion cap.

The tube lock

The London gun trade was initially very reluctant to accept the new percussion lock, principally because its craftsmen had spent their working lives on flintlocks and the change constituted an upheaval in manufacturing. However, in 1816 British gunsmith Joseph Manton took out a patent for a percussion lock in which the fulminate was contained in a tube set in the hammer. The tube could be taken out and reloaded. The disadvantage of the lock was that it was necessary to carry a number of loaded tubes when shooting, as a new one was needed for every shot.

The problem of the volatile fulminate, however, remained unsolved. One solution was to put the fulminate between two paper discs. Another was to make the fulminate into pellets covered with wax or iron oxide. Fulminate was very susceptible to damp and deteriorated quickly, so it had to be carefully stored.

In 1818 Joseph Manton took out another patent, this time for a tube lock in which the fulminate was contained in a copper tube. A section of the tube went directly into the touch-hole, and the hammer fell on

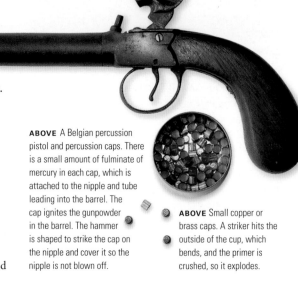

ABOVE A Belgian percussion pistol and percussion caps. There is a small amount of fulminate of mercury in each cap, which is attached to the nipple and tube leading into the barrel. The cap ignites the gunpowder in the barrel. The hammer is shaped to strike the cap on the nipple and cover it so the nipple is not blown off.

ABOVE Small copper or brass caps. A striker hits the outside of the cup, which bends, and the primer is crushed, so it explodes.

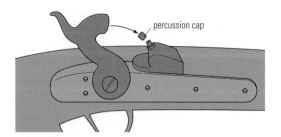

percussion cap

ABOVE Diagram showing the positioning of the percussion cap. The percussion system was relatively straightforward. The hammer struck a nipple on top of which was placed a copper cap containing the priming charge. This exploded on impact and set off the main charge.

the exposed section. This was held in place on a flat pan by a cover and spring. The detonation was very rapid but it tended to blow the tube to fragments – a danger to both the shooter and his immediate neighbours.

The problem of safely housing the fulminate was only solved with the invention of the percussion cap. This was made in the form of a small cylinder shaped like a top hat, which contained the fulminate. The lock was fitted with a nipple which was adjacent to the barrel, and had a channel leading directly into it. The cap fitted over the nipple and was detonated by the hammer when it fell.

Various claims for the invention

The percussion cap was so successful that many of the prominent sportsmen and gunmakers of the day claimed to have invented it. Colonel Hawker wrote an account in *Instructions to Young Sportsmen*, published

in 1830, saying it was his idea and that the gunmaker Joe Manton had altered a pellet-lock gun to his design. Gunmaker Josef Egg claimed to have made a copper cap out of an old penny piece while James Purdey claimed a cap out of the brass ferrule of an umbrella. The most likely inventor, however, is English-born Joshua Shaw (1776–1860), who claimed to have made a series of percussion caps, first in steel, then in pewter, then in copper, in 1816. He went to America in 1817 and was finally granted a patent in 1822. The patent office must have believed his claim because in 1847 Congress awarded him $18,000 for the invention.

The earliest European patent for a percussion cap was granted to François Prelat, a Parisian gunmaker, in 1820 as an addition to another patent issued in 1818. However, as Prelat is known to have copied other gunmakers' discoveries, it is unlikely that he actually invented it, and Joshua Shaw is the most likely inventor.

ABOVE Shooting was to become a fashionable activity. This is a shooting gallery in a smart part of London, *c.*1825–1830.

Adopting the new system

One of the reasons why the percussion cap was slow to be adopted was the composition of the fulminate. The most commonly found priming powder used in early percussion locks was made from potassium chlorate, sulphur and charcoal, along with some other ingredients. The mixture gave very variable results and it also quickly corroded the nipple and the gun barrel.

In 1824 a London chemist took up the manufacture of percussion caps. He based the priming on fulminate of mercury, following a paper published in the *Philosophical Magazine and Journal* by E. Goode Wright, who advocated the use of fulminate of mercury because of its ease of manufacture and relative stability compared with other methods.

The under-hammer lock

The development of the percussion lock led to major changes in the design of firearms. One result was the under-hammer lock. The percussion lock was usually located on the top of the firearm, but on the under-hammer lock it was located on the underside. Under-hammer locks, however, were rare. Their advantage was that during shooting, the flash did not impair vision.

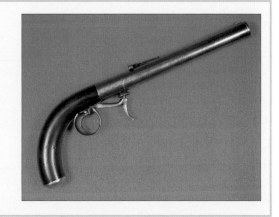

RIGHT Under-hammer box lock percussion pistol, of around 1850 with a turn-off rifled barrel. New designs allowed for a cleaner exterior and greater ease of concealment.

The development of cartridges

As with many handgun developments, there was one overriding criterion. This was the time it took to reload. Soldiers under fire were under pressure to reload as quickly as possible before the enemy was upon them. Sportsmen also needed to reload quickly before their prey escaped. The answer was the cartridge, where all the elements required for firing were contained in a single package.

Loading the gunpowder

In the early days of firearms, gunpowder was loaded into the muzzle of a barrel, followed by cloth and wadding, and then a lead ball. Hunters carried a powder flask, while the military used paper cartridges. The cartridge began as a thick paper tube, which contained both powder and the ball. When it was used with a muzzle loader, the base of the cartridge would be ripped off and the powder poured down through the muzzle into the barrel. The ball would then be inserted and the paper would be shoved down on top of the ball with the use of a ramrod, in order to keep powder and ball in place.

In the days before the flintlock, serpentine powder was used to prime the pan. Later, serpentine was no longer used and the user would prime the pan with some of the powder from the cartridge before pouring the remainder down the barrel. The flintlock

ABOVE 17th-century English powder flasks. The ritual of loading a flintlock musket, or pistol, with powder from a flask and a separate bullet came to an end with the invention of the cartridge.

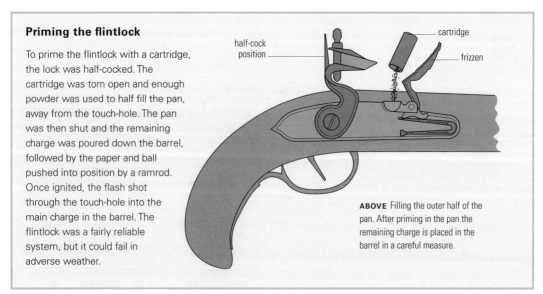

Priming the flintlock

To prime the flintlock with a cartridge, the lock was half-cocked. The cartridge was torn open and enough powder was used to half fill the pan, away from the touch-hole. The pan was then shut and the remaining charge was poured down the barrel, followed by the paper and ball pushed into position by a ramrod. Once ignited, the flash shot through the touch-hole into the main charge in the barrel. The flintlock was a fairly reliable system, but it could fail in adverse weather.

half-cock position

cartridge

frizzen

ABOVE Filling the outer half of the pan. After priming in the pan the remaining charge is placed in the barrel in a careful measure.

RIGHT A Belgian pinfire six-shot revolver from *c.*1865. This revolver was designed to take a metallic cartridge based on Lefaucheaux's design.

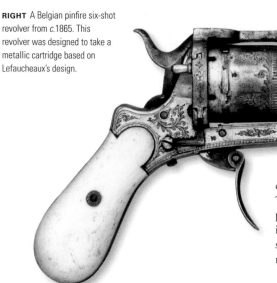

mechanism obviated the need to prime the pan as a separate operation, as a certain amount of powder would run through from the base of the barrel into the covered pan. This system was to prevail during most of the extensive flintlock era, and was mostly reliable as long as the powder in the flash pan stayed dry. This, of course, was not always the case, and on average three out of ten times the flash-pan powder would fail to ignite or the flash pan would ignite without setting off the main charge in the barrel.

After the development of the percussion system by John Forsyth and others, and the parallel development of fulminate of mercury, the next logical step was to contain all the various elements – the ignition, the main charge and the bullet – in one package. There were various experiments with different forms of cartridge, and there were also different schools of thought about the position of the percussion cap in the cartridge.

The pinfire cartridge

Johannes Pauly was a Swiss gunmaker working in Paris. In 1812 he developed and patented a breech-loading firearm and a new kind of cartridge, which was the early development of the pinfire cartridge. These designs were very well thought out, but for various reasons, gunmakers of the time did not follow up his ideas. Work continued in this field, and the pinfire cartridge was invented by a Frenchman called Casimir Lefaucheaux in 1835. The cartridge consisted of a brass case containing a percussion cap and a powder charge. The bullet filled the end of the brass case. A metal pin protruded from the side of the case and when knocked inwards it would ignite the percussion cap. This system proved to be popular and successful, and a number of different weapons were built around it.

One drawback of the Lefaucheaux system was that the protruding pin made the round less convenient to handle than the later rimfire or centrefire cartridges. The pin needed to be aligned correctly in a slot in the chamber, and it could cause the round to detonate if it were accidentally knocked. Lefaucheaux's son Eugene continued to improve both the cartridge and the design of the weapon after his father died in 1852. The system was refined and patented by another Frenchman, Clément Pottet, in 1855, and further work was carried out in London by Colonel Edward Boxer, who patented a centrefire cartridge with a metal base and coiled brass body.

BELOW The kit of an American Civil War cavalryman included French pinfire pistols and a cartridge case.

Paris gunsmiths such as Lefrancheaux were influential in the development of composite cartridges comprising paper and brass. A similar version was successfully used by gunsmith Johann Nikolaus van Dreyse for a needle gun – so named because of the extended pin that ran through the centre of the bolt to ignite the percussion cap. The most notable development was the placement of the percussion cap in the centre of the base of the brass cartridge, which is the same system used in modern centrefire cartridges.

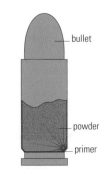

ABOVE The first rimfire cartridge was essentially just a percussion cap with a projectile pressed in the front, and a rim to hold it securely in the chamber.

The rimfire cartridge

The rimfire cartridge developed in the 1830s, although it was not started commercially until 1845. The rimfire cartridge was essentially a large percussion cap which contained not only the priming compound, but also the propellant powder and bullet. One disadvantage of the rimfire system was that the cartridge is rendered useless by the impact of the hammer and it could not be re-used (unlike other cartridges which could be reused by replacing the primer, gunpowder and bullet). It also had a shorter range. Due to the thin case they are limited to light, low pressure calibres (calibres relate the diameter of the inside of the barrel). If the calibre measurement is in inches then the calibre is quoted as decimal of an inch, so a gun with a diameter of 0.22in is a .22 or "twenty-two calibre". In the past, rimfire calibres up to .44 were common. Today's rimfires are of calibre .22 (around 5.5mm) or smaller.

The centrefire cartridge

In 1854 Charles Lancaster invented a cartridge which was sealed in a brass case. The brass at the base was thin enough for the pin to activate the percussion cap contained in the base of the cartridge.

ABOVE Pinfire pistols were used during the American Civil War (1861–65), although they were not always highly regarded because of their low power compared with other percussion revolvers, such as Colts, which were widely used on the battlefield.

This was a centrefire cartridge, which proved to be the most successful cartridge type and is one of the most commonly used for all large-calibre weapons today. There is some discussion as to the original patent for the centrefire cartridge, which is in itself an indication of the significance of the invention. It is said to have been invented in 1861 by the English gunmaker G. H. Daw, but Clément Pottet of Paris is also mentioned, and a London firm, the Eley Brothers, successfully won the patent.

In 1867 an Eley–Boxer centrefire cartridge was adopted by the British War Office for use in Enfield rifles. The original Eley–Boxer cartridge was made of solid brass. In 1870, Colonel Hiram Berdan of the US Ordnance Department developed a bottle-shaped brass cartridge with a cap and chamber at the base and flash holes drilled on either side. Modern cartridges are made from a brass and metal alloy. Centrefires today also are generally capable of higher-powered loads than rimfire.

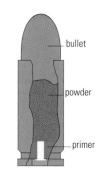

ABOVE A centrefire cartridge has the primer located in the centre of the base. Centrefires are generally capable of high-powered loads.

The advantages of a metal cartridge

The beauty of a metal case for some types of cartridges was that it expanded under the heat of the explosion and thus created a seal in the chamber. This increased the energy of the bullet towards the muzzle of the gun. It also prevented the gun user from being injured by escaping high-pressure gases as the trigger was pulled. The brass case of a centrefire cartridge could also be re-used once it had been ejected from the chamber. It had to be precisely manufactured to suit the dimensions of the particular gun for which it was intended. This entailed precision machining to fine tolerances of the parts integral to the gun and the cartridge itself. Different cartridges would be designed to carry different loads, and it would be important to match the right cartridge load to the pistol, otherwise the pistol would be liable to explode when it was fired. Cartridges are manufactured and named in imperial or metric sizes.

The needle gun

Johann Nikolaus Dreyse was a Prussian gunsmith who invented a military breech-loading needle gun. The system involved the use of a bolt which was used to open and close the breech, through which the cartridge was loaded. The bolt was locked by pulling it down in front of a lug in the frame. The "needle" was an extended firing pin that passed through the centre of the bolt. When released by the trigger, the needle would fly forward, pass through the paper cartridge and ignite a percussion cap at the base of the bullet. The cartridge used in this system was made up of four separate elements: the paper case itself, the powder charge, the priming cap and the bullet.

The primer was placed between the bullet and the main powder charge so that the needle passed through the powder charge before igniting the primer.

Although it was a clever design and was highly effective for its time, the needle gun also had some inherent weaknesses. The fact that the needle had to pass through the bolt meant that, unless the parts were extremely well fitted and sealed, a certain amount of gas would leak through the bolt. Also, as the needle was present when the explosion took place, it was quickly subjected to wear and tear and would sometimes break.

ABOVE German needle-fire bolt-action military rifle from around 1851. The effectiveness of the needle and cartridge system was soon shown on the battlefield when the Prussian army succeeded in producing a more rapid rate of fire than its enemies. Other countries, such as France, soon caught up.

Manufacture, proof and trade

The term "lock, stock and barrel" is a reminder that the firearm was made up of several different components, each of which required different materials and different craftsmanship skills. The inherent and increasing complexity of the firearm was to pose a growing technological challenge to gunmakers, while the organization of the different trades was to become both a stimulus to and a sign of the growing process of industrialization, first in Europe and then elsewhere.

Craftsmanship

The manufacture of a firearm was a complex process and involved a number of different specialist trades relevant to the different parts of the firearm. The 16th-century origins of the firearm manufacturer Beretta, for example, lie in precision forging of steel barrels. As firearms became more complex, they not only called on the skills of metalworkers and forgers but also of woodworkers to develop the stock and experts in precision mechanisms to create the trigger and lock systems.

Each part of the lock mechanism for a flintlock pistol would have been hand-made. The trigger, powder pan, springs and screws would have been individually forged and filed.

ABOVE A gun is assembled at the Beretta gun factory in Italy. Craftsmanship skills would prove to be compatible with the advent of the machine age, and successful firearms manufacturers would find the right balance between the two.

Although early stocks for such weapons as the arquebus or matchlock were fairly rudimentary, later stocks for flintlocks, percussion weapons and breech-loaders were an integral part of the weapon. These needed to be worked with skill so that the metal parts fitted correctly and in alignment with each other, and so that the stock itself was strong enough to both support the weapon and absorb recoil.

Stocks were often made of well-seasoned hardwoods such as English walnut, although this did not necessarily always come from England. In addition to their utilitarian value, stocks were sometimes engraved to underline their fine workmanship. The barrel itself was often forged from a flat wrought-iron bar. Metalworking skills that included heating, hammering and boring would result in a smooth-bore interior. A separate process was used for rifling.

The quality of weapons depended to a large extent on developments in metallurgy. Weapons could not be made to fine tolerances if the metal could not be worked precisely. Due to the high pressures involved when a weapon was fired, this was of some significance.

The art of gunmaking had to take many factors into account. It needed to be fitted to its purpose, to have the right balance and "feel", and such details as the quality of the trigger pull would also be noticed. The craftsmanship of the gun evolved in the manufacture of finely weighted barrels, the creation of smooth-functioning locks and triggers and the sculpting of a fine stock. Alongside this individual craftsmanship came the advent of the machine age. Machines were designed to produce parts in greater volumes, though without the individual touch. The two processes could be successfully combined by bespoke gunmakers while volume manufacturers could claim that their products were often on a par with their handmade competitors.

The Birmingham Gun Quarter

The different skills required in the manufacture of a firearm were part and parcel of the process of industrialization in England. Different craftsmen focused on their individual products and they were brought together by the gun manufacturers to assemble the finished product. It made sense

to keep the different trades as close together as possible in order to cut down on time and expense and as a result areas such as Birmingham's Gun Quarter became recognized as specialist areas for gunmaking. Not all of the parts were produced in the Gun Quarter, however. Locks, for example, often came from elsewhere.

In 1767 Birmingham in England had 35 gun and pistol manufacturers, 8 gun barrel manufacturers and filers, 5 gun barrel polishers and finishers, 11 gun lock manufacturers, forgers and finishers, and three gun-swivel and stock makers. The different parts were often sent to London to be made up by "fabricators" who assembled and sold the weapons.

LEFT Manufacture of arms around 1862. The organization of craftsmen and machinery to satisfy the demand for firearms was a significant stage in the Industrial Revolution.

Proving

The Birmingham Proof House is evidence of another important stage in gunmaking – the proving (testing) of barrels. The gun barrel would have a charge fired through it that was significantly more powerful than the one it was designed to fire. If there was any sign of weakness, such as distortion or cracking, the gun would fail the test. Otherwise, it would be awarded an official proof mark to prove its worthiness. The Birmingham Proof House was established in 1813 by Act of Parliament, and it became illegal to trade in arms without the award of a proof mark.

Due to the particular nature of gunmaking, it both thrived upon and contributed to the greater sophistication of the nascent industrial age. With the growth of large cities and, in the case of England, with labourers beginning to leave the land to seek work in the towns and cities, there was great scope for the development of a thriving gun trade.

The United States

In colonial America, gunsmiths also sprang up, though initially many of these would either assemble or repair firearms imported from England and elsewhere. In due course, the United States would be the pioneer of some of the most important small-arms designs of all time and would develop its own thriving gun trade.

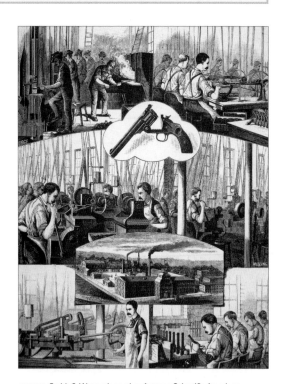

ABOVE Smith & Wesson's revolver factory: *Scientific American* 24 January, 1880. Like Colt, Smith & Wesson soon became a major industrial concern with growing markets both at home and overseas.

Duelling pistols

As firearms became more accurate, they were used for hunting and for sport. In due course, the handgun in particular came to be used in duelling, a practice that had its roots in a warped code of honour. While an "honourable scar" might be obtained from the slash of a sword, firearms were inherently less difficult to control and the results were almost invariably fatal. In order to meet the ongoing demand, specialist manufacturers began to emerge who prided themselves in the sophistication and accuracy of their products.

A code of honour

Duelling has had a long and undistinguished history that pre-dates the medieval era. The throwing down of a gauntlet was the traditional challenge that would be followed in the pre-firearms era by a fight with swords. In England there was such a thing as a judicial duel, dating from the 11th century, which was only abolished by English law in 1819.

The nature of duelling, however, is such that it tells us less about the law of the state than it does about individual codes of honour. So precarious was the notion of honour that the offended person was "honour bound" to demand satisfaction for even the smallest slight, or risk a disastrous decline in his status. Similarly, the accused would either have to accept the challenge or risk being branded a coward. It was the direct opposite of the Christian teaching of "turn the other cheek".

ABOVE A matched pair of flintlock duelling pistols made by London gunmakers in 1805. Duelling pistols were designed to be identical in every way to give both contestants an even chance.

The use of pistols for duelling became more common from about the middle of the 18th century and, in order to ensure that the duellists should have an equal chance, the pistols were often designed as matching pairs and presented in a box. In some cases the box also contained miniature pistols to be used by the "seconds" (trusted representatives of each party who may stand in for the dueller). This would allow the seconds to defend themselves or to intervene if there should be any unorthodox behaviour.

Early makers of duelling pistols

The first set of duelling pistols was probably made in England, to very high specifications. As duelling was supposedly the preserve of "gentlemen", it was appropriate that the design and finish of duelling pistols should be suitably lavish, and this was particularly the case for French pistols. It was also important that the pistols should be both reliable and accurate. In the days of flintlocks, great care was taken to minimize the chances of a misfire by careful loading.

There were a number of pistol makers in both Europe and America who acquired considerable reputations for the quality of their products. These included Durs Egg, John Manton, John Twigg and Robert Wogdon of England; Gastinne Renette, Le Page and Nicolas Beutet of France; Auguste Francotte of Belgium; Continner of Vienna; and Simeon North, James Haslett and R. Constable of America.

Early duelling pistols tended to be lightweight designs, but heavier designs began to emerge. Part of the extra weight was in the barrel, which was designed to reduce recoil and improve accuracy. As with many flintlock firearms, duelling pistols were often converted to the percussion system, which radically increased their reliability.

Code Duello

Duelling was controlled by elaborate codes, which included the French Renaissance code, and an Irish code agreed at the Clonmel Summer Assizes in 1777 which was to some extent based on the French code. The Irish code had over 25 rules that stipulated how a duel should be managed and what constituted satisfaction for an insult.

The basis of any code was that, once an insult had been made, a challenge would be given and a place agreed where the duel could take place. Normally the contestants would bring seconds as well as a physician. The seconds might try to arrange a reconciliation between the parties and, if an apology was accepted, the duel would be called off.

The duel would be carried out on a piece of measured ground and details such as the number of paces between the contestants would be agreed.

ABOVE Pistols were increasingly used for duelling from the mid-18th century.

RIGHT Normally duels were carefully stage-managed. The "quick draw" from a holster belonged to the Wild West.

Sometimes the contestants would "delope", which meant firing into the air so as not to wound the other party. This practice was, however, forbidden in the Irish code.

Famous duels

There have been a number of notable duels through the ages. In 1598 the playwright Ben Jonson killed the actor Gabriel Spenser. In 1796 William Pitt the Younger had a duel with George Tierney. In 1809 George Canning had a duel with Lord Castlereagh and, in 1829, the Duke of Wellington had a duel with the 10th Earl of Winchelsea.

The duel between the duke and the earl followed public criticisms of the duke by the earl. The two met in Battersea Fields and the duke fired wide. Winchelsea returned the compliment, after which he agreed to publicly retract his accusations. The duel damaged the reputation of the Duke of Wellington and underlined public distaste for the practice of duelling. It is a good indication of the level to which the code of honour had become ingrained that, in two countries where the practice of duelling had been declared illegal, the Prime Minister of Great Britain and two leading American politicians should use the duel to seek their satisfaction. At the height of its notoriety, hundreds of duels per year were fought between less well known individuals.

The Hamilton–Burr duel

In 1804 the United States Vice-President, Aaron Burr, fought a duel with the former United States Treasury Secretary, Alexander Hamilton, in which Hamilton was killed. By the time the Hamilton-Burr duel was fought, duelling had been declared illegal in New York, so the two men and their accomplices secretly crossed the Hudson River to a remote rocky ledge in New Jersey.

It is said that Hamilton may have deliberately aimed wide with his first shot, though another view is that, the bullet from Hamilton's gun was launched too early and went high. Burr is also said to have tried a non-fatal shot, but it did enough damage to Hamilton's internal organs to kill him.

ABOVE The 1804 duel between Aaron Burr and Alexander Hamilton in which Hamilton was killed. This duel was fought despite legislation in New York that prohibited duelling.

Eastern pistols and handguns

Firearms from the East have a rich and distinctive history. They are recognizable, both for their unique shapes and also for the characteristically rich decoration that distinguish them from more sober European products. Although some of these firearms were decorated European derivatives, many fine and original products were also developed in the East.

A slow development

Firearms were not developed in the East with the same energy and purpose as they were in Europe. The military tactics that flowed from the appearance of firearms were slow to develop. Codes of honour and concepts of how a warrior should behave meant that the sword was often preferred to the firearm. Moreover, unlike Europe, where a process of industrialization complemented firearms development, the manufacture and maintenance of firearms in the East was not as well developed.

However, it is almost certain that gunpowder was invented in the East, and the knowledge of its secrets are said to have been transmitted to Europe via Iran and India, where they are likely to have been recorded in Sanskrit, and then to the Maghrib (the West), which included Moorish Spain. In the 13th century the Arab alchemist Hasan al-Rammah wrote a description of how to make gunpowder, and Roger Bacon and Albertus Magnus visited the Spanish-based centres of learning to find out about it.

Despite its Eastern origins, the conservatism of the Ottoman Empire was such that little was done to take the next logical step – to develop the means by which the gunpowder could be used. It is said, however, that the Arabs fired a missile from a sealed tube of bamboo, filled with black powder, and the Chinese are believed to have developed one of the first metal handguns.

By the time of the Siege of Algeciras in southern Spain by Alfonso XI in 1343, the Christians discovered to their cost that the Muslims were armed with cannon.

ABOVE This late 18th-century Algerian pistol with coral inlaid stock formed part of the presentation by the Algerian Ambassador to King George III of England in the early 19th century.

The Balkans

The transfer of firearms technology from the West to the East largely took place via the Balkans. The Venetians are said to have taken firearms to the Balkans in 1352 and the Ottomans used firearms to defeat Christian crusaders at battles such as the Battle of Kosovo (1448).

Due to the ongoing threat of Muslim expansion, a Papal Bull of 1444 banned the trade of firearms with the Ottomans, but this and other bulls do not appear to have had much effect. After the dissolution of the monasteries in England in 1541, lead and bronze from the destroyed buildings were sold to the Turks, some of which may have been used for making weapons and ammunition.

The elite Ottoman corps known as the janissaries were the soldiers most likely to be equipped with firearms. The janissaries included a large proportion of young, originally Christian men from the Balkans who had converted to Islam, and they largely replaced the traditional cavalry, the spahis, as footsoldiers.

By the beginning of the 16th century, handguns were widely in use in the Ottoman Empire, but many of these were of European origin or were inspired by European designs and it is said that European-sourced firearms were preferred to the local variety by those who could afford them. It was soon discovered that when they malfunctioned there was little or no native expertise or backup from armourers to put them right.

ABOVE Siege of Belgrade, 1521, from *The Military Campaigns of Suleyman I*. The exchange between East and West in the Balkans would also include the transfer of knowledge about firearms.

A pistol might become little more than an ornate and expensive club. As in Europe, homemade firearms were made by different craftsmen and there was a system of testing and approval among the Turks, with the appropriate mark made on the pistol once it had been passed.

In the 18th and 19th centuries firearms continued to be imported into the Ottoman Empire in ever-greater numbers, and it provided a valuable market as the Industrial Revolution gathered pace. English firms such as the Birmingham Small Arms Company won substantial orders from the Ottoman Government.

The Maghrib

As European nations such as Portugal pushed down the west African coast in their quest for the route to India, and as France and Spain extended their influence in North Africa, the development of firearms in the Maghrib was duly influenced by European designs. Again, innate conservatism led to the retention of certain archaisms in design, and local influence was seen largely in the style of decoration,

which might include gold and silver wire or gold and silver damascene. In some parts, notably Algeria, coral was also widely used. Again, different craftsmen would produce parts of the firearm, and often these were manufactured to a high standard.

The Caucasus

Like the Balkans, the Caucasus is a link between Europe and Asia, and the region has its own tradition of weapon-making which speaks of a high level of craftsmanship and enjoys a high reputation. Typical pistols from the region from the beginning of the 19th century would have ivory ball butts and black, leather-covered stocks. Decoration with niello (an alloy used as an inlay in a design) was common.

Japan

Firearms were first introduced into Japan in 1542 by the Portuguese. The local warlords were keen to trade for more arms, and they also ordered local swordsmiths to copy the weapons. By 1575, at the Battle of Nagashino, Japanese battle tactics had changed to suit the introduction of firearms. By 1587, however, it was thought they might give too much power to the peasantry. Under the military dictatorship of the Tokugawa Shogunate, which lasted until 1868, firearms were believed to have been suppressed as part of a wider plan to maintain stability, but this was mostly notional, and on the whole they continued to be used.

ABOVE *Geijutsu Hideu Hue* (Accomplishments in the Secret Arts) by Japanese Ohmori Sakou, illustrated by Kuniyoshi, reveals the Japanese use of firearms in the mid-19th century.

The art of the gunmaker

Early firearms were sophisticated pieces of workmanship and the owners of these treasured possessions were invariably wealthy. Alongside the craftsmanship required to produce a firearm there grew the craft of decoration. The more complex the firearm was, the more elaborate was the decoration. The level of decoration, however, varied between different countries, with France producing some of the most lavish handguns and England the most sober. Gunmaking centres created their own styles.

Early decorated handguns

Even in their very earliest forms, firearms were sometimes decorated. The barrel of the late 14th-century cast-bronze handgun from Morko is roughly engraved with religious texts on the barrel and has the bust of a bearded man adjacent to the touch-hole. Bronze barrels were made in late medieval times, by founders used to producing decorated bronze wares, so it was no wonder the barrels were often ornate.

The earliest examples of decorated handguns date from the 1530s, including wheel-locks made for the Emperor Charles V (1519–56). The number of finely decorated wheel-lock guns far outnumbers those with matchlock mechanisms – this expensive, complicated mechanism was deemed to deserve rich decoration. Craftsmen from various traditions, specializing in decoration, therefore found themselves practising their skills on firearms. Barrels were chiselled and gilded, and wooden stocks inlaid with engraved staghorn.

ABOVE A pair of pistols by N. Boulet made for Murat in 1805, Musée de L'Armée, Paris. France produced some of the most lavishly decorated firearms, with French gunmakers dominating the trade in the 17th and early 18th centuries. It was a celebration of craftsmanship.

There has always been a link between the craftsmen who worked on gun-stocks and those who worked on small items of furniture such as cabinets and desks, because of the use of materials such as staghorn and ebony which were widely used for inlaid wood in the 16th and 17th centuries. The craftsmen who decorated high-quality firearms were careful to avoid techniques that would weaken either the stock or the barrel. For this reason, in the 16th century, etched decoration was always preferred to chiselling, and inlaid decoration to carving.

Sources of ornament

It seems from surviving examples that highly decorated firearms first began to be produced in quantity in the first quarter of the 16th century. Elaborate decoration was based upon the engraved sheets of designs which circulated in the workshops of different gunmakers. The craftsmen employed on the decoration of a gun would reflect elements of an engraved design, fitting them into the space available on a gun-stock or barrel. It is possible therefore to see the influence of artists such as German artist Virgil Solis and French goldsmith Etienne Delaune on firearms decoration.

A finely decorated firearm was the work of several craftsmen. The wooden stock was made by the stock-maker, the barrel maker produced the barrel, and the lock was made by yet another craftsman. Engraved designs made specifically for firearms were not apparent before the last quarter of the 16th century. After this period they become common.

It is from the 17th and 18th centuries that most surviving pattern books date, and it is possible to relate engraved and chiselled work from these periods to the pattern books of particular artists. Among the more notable are François Marcou (c.1657), Jean

Berain (*c*.1650) and De Lacollombe (*c*.1730). All three artists were French, but as French gunmakers dominated the trade in the 17th and early 18th centuries this is not surprising.

Gunmaking centres

Various gunmaking centres had their own particular decorative styles. Northern Italy, especially Brescia, used pierced steel and chiselled work; French gunmakers used chased gold; and the gunmakers of Tula in Russia used blued steel overlaid with tiny flowers in gold. By contrast, the best English makers used decoration very sparsely – the maker's name and some simple engraved scroll-work is often the only ornament. Silver mounts are frequently found on firearms, usually cast from moulds and applied to the stock. For the highest-quality work cast gold was used. The neo-classical presentation pistols by the French gunmaker Boutet are often mounted in gold. Five pistols made for the 19th-century exhibitions of 1851 and 1882 also by French gunmakers, were highly decorative and cannot have been intended for use. Some fine early Colt revolvers are richly engraved, but had the overall sense of design of the earlier periods.

Decorative techniques

A variety of techniques was used over the centuries to decorate firearms. The metal parts such as the barrel, lock, butt-cap and mounts were etched and gilded, damascened (inlaid) in gold and silver and chiselled. Some of the first chiselled work was done by Daniel and Emmanuel Sadeler, who worked for the Elector of Bavaria in the 1620s. The designs were usually based on the engravings of Etienne Delaure and were in low relief against a gold background, the relief decoration being coloured dark blue to make a vivid colour contrast. The stocks of pistols were inlaid with a variety of metals, especially brass and silver.

The subjects were drawn from mythology or from the chase. Firearms were widely used for hunting, especially in the 16th and 17th centuries.

Gunmakers followed contemporary ornamental styles extremely closely. French Rococo and Neo-classical styles are all found on firearms and there are even pistols made for presentation in the 19th century that are decorated in the style of the gothic revival, the stocks being set with medieval knights in architectural niches of carved ivory.

BELOW Dutch gunsmiths at work, *c*.1695.

Combined weapons

The somewhat unpredictable nature of early firearms, added to the elaborate and lengthy process of reloading, gave rise to the notion that the firearm should be combined with an edged weapon or another form of defence.

Early inventions

The idea of combining the mechanism of a firearm with another weapon was thought of as early as the 15th century. Some of the earliest wheel-locks were found on three crossbows in the Arsenale at Venice, dating from about 1510. Another interesting group of combined weapons are the gun-shields from the armoury of Henry VIII of England, many of which are preserved in the Tower of London and Windsor Castle, in England. Supplied in 1544 by an Italian merchant in Ravenna, these consisted of a steel-faced circular shield, fitted in the centre with a short, heavy, breech-loading pistol operating on the matchlock principle.

In addition to guns combined with maces and axes, firearms were also combined with a dagger. The wheel-lock mechanism was mounted on the face of the blade and the barrel was formed of the central section of the blade. The dagger became a gun when the tip of the blade was pulled out to reveal the barrel.

Different gunmaking centres produced their own distinctive combination arms. One such was the *fokos*, an axe with a small blade, mounted on an elaborately inlaid wooden shaft, on which is set a flintlock mechanism. The *fokos* was a speciality of the gunmakers of Teschen in Silesia and was popular in the late 17th century, mainly in Eastern Europe.

Hunting swords

Where combined weapons were most widely used was in the hunting-field. Hunting was a most important occupation, especially before the 19th century. Elaborate costumes and equipment were designed specifically for the chase. Included in this equipment were combined arms. These included spears combined with guns and, in particular, sword gun combinations. There are a number of short hunting swords dating from the 18th century which have short barrels mounted on the blade with a small flintlock set near

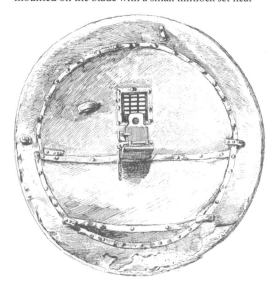

ABOVE Gunshield with a short barrelled breech-loading matchlock gun in its centre. Stored in the Tower of London since the 16th century, the shield was mentioned in an inventory dated 1547.

ABOVE Rear view of a 15th-century English shield with a gun. The small trellised window would have been used for aiming. Such designs were impractical and short-lived.

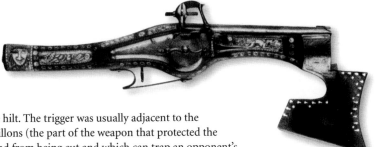

LEFT A German 17th-century wheel-lock pistol with battleaxe. Apart from the sword or knife, most combination weapons such as these were clumsy and inefficient and have never been very successful in use.

the hilt. The trigger was usually adjacent to the quillons (the part of the weapon that protected the hand from being cut and which can trap an opponent's blade). Mention should also be made of the series of flintlock firearms produced at the end of the 18th century that had a bayonet folded under the barrel which snapped forward into place when the trigger guard was pulled back.

Combining with furniture

One of the most unusual examples of a firearm combined with a piece of furniture is a 19th-century steel chest made to contain valuables which has four percussion pistols mounted inside. If the lid is opened without setting a special catch, the pistols fire. An examination of the barrels of the steel chest indicated that all four pistols had been fired at some time.

Other combinations

Pistols have been combined with keys, with knives, forks and spoons, and with truncheons. An inventor from Barnstaple in Devon, England, patented a metal truncheon with a percussion lock in 1823. An American patent of 1837 obtained by George Elgin of Georgia was a single-shot percussion pistol combined with the blade of a Bowie knife (a stout hunting knife). These were made in some quantity and were clearly very popular. In Sheffield a knife-pistol known as the "self protector" was made by the cutlery firm of Unwin and Rogers. These were made from about 1845 and were large pen-knives set with a barrel.

Wheel-lock combination axe and pistol

This was possibly a novelty or curiosity weapon, which fired from five barrels from the axe-edge end of the weapon. The muzzles of these five barrels were concealed by a hinged cover which formed the edge of the axe-blade at the right. The topmost barrel was ignited by a matchlock (not visible) fitted on the left side – the mechanism was concealed by a brass plate. The second barrel was visible from the right side of the weapon and ignited by a wheel-lock. This mechanism occupied most of the outer surface of the axe-head on the right side. The gunpowder in the pan was ignited by the pyrites held in the jaw. There was a tubular extension to the pan of the wheel-lock to hold a length of match which would be ignited by the flash of the priming and then withdrawn to ignite the remaining three barrels. The remaining three barrels were ignited by the hand-held match, as was the sixth barrel forming the handle or shaft.

RIGHT This unusual weapon is believed to be Iberian or German, dating from the early 17th century.

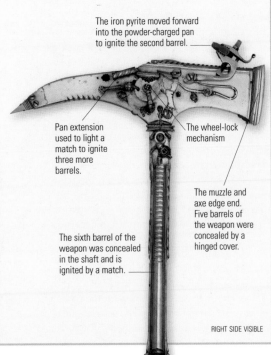

The iron pyrite moved forward into the powder-charged pan to ignite the second barrel.

Pan extension used to light a match to ignite three more barrels.

The wheel-lock mechanism

The muzzle and axe edge end. Five barrels of the weapon were concealed by a hinged cover.

The sixth barrel of the weapon was concealed in the shaft and is ignited by a match.

RIGHT SIDE VISIBLE

"The gun that won the West"

The handgun most widely associated with the opening up of the American West in the mid and late 19th century, and the books and films that have been based on this period, will always be the Colt revolver in its various forms. If this was the gun used by the "good guy" in a Hollywood Western movie, the weapon of choice of the villain was the pocket sized Deringer pistol. In the 1890s Smith & Wesson, a company with a fine pedigree in small-arms design would produce the revolver that would replace the Colt for the US Navy and Army.

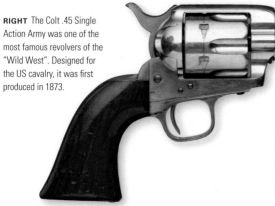

RIGHT The Colt .45 Single Action Army was one of the most famous revolvers of the "Wild West". Designed for the US cavalry, it was first produced in 1873.

Samuel Colt's innovation

The style of revolver that was popular in the 1840s and 1850s had six bored holes in the barrel. It could be loaded with six shots in one go, which could be fired one after the other. It continued to be used until the 1860s in both Europe and the United States. In 1835 Samuel Colt took out a patent for a weapon with a rotating cylinder. He established the standard of a cylinder with multiple chambers, each of which successively locked in position behind the barrel and was discharged by pressure on the trigger. In Colt's early revolvers, the cylinder revolved as the hammer was cocked manually. Later revolvers, in which the hammer was cocked and the cylinder revolved as the trigger was pulled, were developed soon afterwards.

The .44/40 Colt

Colt initially produced the same gun in a variety of calibres and barrel lengths, ranging from a stubby 4.75in/120mm up to 12in/305mm for the civilian market. The .44/40 Colt was particularly popular because the ammunition was compatible with that of the Winchester carbine (known as "the gun that won

the West"). The US Army bought about 37,000 Artillery and Cavalry Colt pistols between 1873 and 1893. Commercial production continued, but ceased in 1941. The Colt model 1873 revolver, known as the Colt Single Action, Peacemaker or Frontier was developed by 1872, based on the patents granted to Charles B. Richards and W. Mason. In 1873, the US Army adopted this revolver along with its black powder centrefire .45 Long Colt cartridge, and issued the Army or Cavalry model with two different lengths of barrel.

The advent of films and television Westerns in the 1950s prompted the Colt company to bring the Single Action back into production for the commercial market in 1956. The 1957 film *Gunfight at the O.K. Corral*, depicting a fight on a chilly day in 1881, is an interesting example of the realities of gunfights in the Wild West. Handguns were notoriously difficult to fire at close range. The accepted wisdom is the classic pose of a gun at arm's length and shoulder height. The quick-firing hero fanning the hammer of his Colt as he "shoots from the hip" and hits his adversaries with lethal shots is pure Hollywood.

Deringer pocket pistols

At the far end of the scale from the Colt revolver was the little single-shot pocket pistol developed by Henry Deringer of Philadelphia. Versions of the Deringer were made by numerous gunsmiths, including Colt. It achieved its greatest popularity in the United States in the years leading up to the Civil War. It was a favourite of men and women who wanted a compact, easily concealed firearm for personal defence, and so 15,000 were produced between 1850 and 1866.

The Colt factory

Samuel Colt, the son of a factory owner, was born in Hartford, Connecticut, on 19 July 1814. After leaving school, Colt worked at his father's textile mill. He was fascinated with machinery and spent his spare time disassembling and reassembling his father's guns. When only 16, he designed a fireworks display for the Fourth of July celebrations. Unfortunately, the school building burned to the ground. Fearing expulsion, Colt ran away to sea in 1829. While at sea Colt got the idea of designing a handgun with a revolving cylinder containing several bullets which could be fired through a single barrel. On his return he continued his work and the five-shot revolver was patented in the mid-1830s. In 1836 Colt established his Patent Arms Manufacturing company in Paterson, New Jersey.

However, sales were generally slow, and in 1842 he was forced to close down his factory. In 1847 Colt designed a six-shot Walker revolver. Soon afterwards the US government ordered 1,000 revolvers for use in the Mexican War. Colt was now in a position to establish a new factory in Hartford, Connecticut. By 1856 Colt had the largest private arms' manufacturing facility in the world. Colt, who skilfully promoted the gun culture that endures in America to this day, died on 14 January 1862. The Colt industry thrived and its handguns would arm the US Armed Forces and Police for many years.

LEFT The inventor and industrialist Samuel Colt. By 1856 Colt owned the largest private arms manufacturing facility in the world.

Although the .41 Deringer pistol was somewhat limited by its single-shot capacity, its light weight and small size gave it a distinct advantage over bulkier weapons like the 1873 Colt. To compensate for the single shot, many people simply carried two pistols. The Deringer became notorious when it was used in a number of Californian murders in the 1850s, as well as its later use in the assassination of President Lincoln. In Hollywood films about the West the little pistol would be used by sneaky gamblers, though in John Ford's 1939 classic film *Stagecoach* the southern gambler produced a Deringer as an act of mercy to kill the wife of a US Cavalry officer and spare her from capture by Apache Indians.

The test of war

In 1892 the US Army retired the Colt .45 Single Action Army Model revolver and decided to re-equip with a more modern revolver. They selected a .38 revolver built on a robust frame and both Colt and Smith & Wesson made versions of these firearms. In 1889 the US Navy adopted the revolver, followed by the US Marine Corps in 1905.

However it was during the Philippine Insurrection of 1899–1900 that showed that the .38in cartridge was not powerful enough. To rectify this, in 1908 a .38 Special cartridge was developed and the .38 Army Special revolver. Though it was superseded by a .45 model of similar design in 1909 the Special remained in production until 1928 for the commercial market and police, with some 220,000 being made.

LEFT *Gunfight at the O.K. Corral,* 1881, as depicted in a painting by Terence Cuneo. It gives a good idea of the close range at which many of these shootouts took place. In this gunfight one participant – John "Doc" Holliday – used a shotgun that he had kept hidden under his coat.

Innovations

The final years of the 19th century saw innovations in the way pistols and revolvers were designed. Among them were firearms made by Leopold Gasser who operated factories across Europe, which reputedly turned out 100,000 revolvers annually in the 1880s and 1890s. In 1907 the Austrian Roth-Steyr self-loader was the first such weapon to be adopted for military service. The huge British Gabbet-Fairfax Mars semi-automatic (also called self-loading) pistol, though novel, was destined never to be used by the British Army.

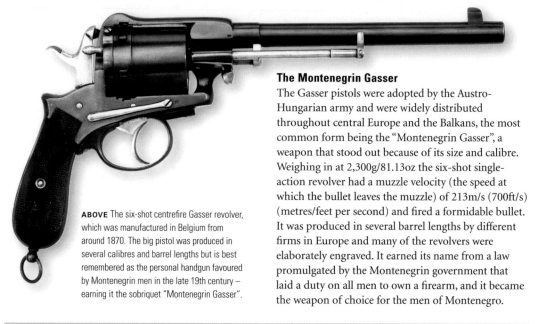

ABOVE The six-shot centrefire Gasser revolver, which was manufactured in Belgium from around 1870. The big pistol was produced in several calibres and barrel lengths but is best remembered as the personal handgun favoured by Montenegrin men in the late 19th century – earning it the sobriquet "Montenegrin Gasser".

The Montenegrin Gasser

The Gasser pistols were adopted by the Austro-Hungarian army and were widely distributed throughout central Europe and the Balkans, the most common form being the "Montenegrin Gasser", a weapon that stood out because of its size and calibre. Weighing in at 2,300g/81.13oz the six-shot single-action revolver had a muzzle velocity (the speed at which the bullet leaves the muzzle) of 213m/s (700ft/s) (metres/feet per second) and fired a formidable bullet. It was produced in several barrel lengths by different firms in Europe and many of the revolvers were elaborately engraved. It earned its name from a law promulgated by the Montenegrin government that laid a duty on all men to own a firearm, and it became the weapon of choice for the men of Montenegro.

The single- and double-action revolver

A revolver works by having several firing chambers arranged in a circle in a cylindrical block. These are brought into alignment with the firing mechanism and barrel one at a time. In early revolvers, a shooter had to pull the hammer back before each shot and then pull the trigger to release the hammer. These are called single-action revolvers.

An innovation in lock-work arrived in the 1850s, which allowed firing by single- or double-action. The hammer could either be cocked and then released by a light trigger pressure, or the trigger could be pulled right back in a "long pull" action to rotate the cylinder and raise and release the hammer. This meant it did not need to be manually cocked between shots. Double-action is used on most present day revolvers. The first revolvers used gunpowder, balls and caps like the early percussion-cap pistols. The shooter would load each of the six chambers in the cylinder with gunpowder and a projectile, and fire using the percussion caps. In the 1870s, these early models were replaced by revolvers that used bullet cartridges instead of gunpowder and caps. Revolvers passed through the pinfire and rimfire stages to use the centrefire metallic cartridge case. The breech-loading revolver has a gate in the rear part of the frame (which backs up the cylinder) for loading and ejecting.

Young cannon

In the world of big self-loaders, the British Gabbet-Fairfax Mars reigned supreme. Different marks designed by H.W. Gabbet-Fairfax could fire 8.5mm, 9mm or .45 calibre ammunition. However, it was the .45 that really had a punch. It had a muzzle velocity of 381m/s (1,250ft/s) while the 8.5mm produced an incredible 533m/s (1,750ft/s). It was rejected by the British War Office as a service weapon because of the requirement for special ammunition and the excessive recoil caused by the powerful cartridges and complex long recoil mechanism, which did not lend itself to cost-effective production. Another drawback of the design was that the cases were ejected out of the back of the pistol directly into the face of the firer. The design did not prove to be a commercial success. It is not known exactly how many Mars pistols were manufactured, most estimates being around 70, although one pistol is known with a serial number of 195. It is an indication of the rarity of these weapons that the starting price for one that came up at auction in 2006 was set between £8,000 and £12,000.

ABOVE Different designs of the Gabbet-Fairfax Mars could fire 8.5mm, 9mm or .45 ammunition. The recoil force from the .45 round was such that the firer found his arm jerked vertically upright on firing.

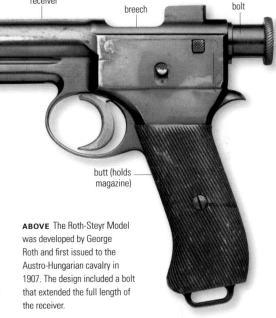

receiver | breech | bolt

butt (holds magazine)

ABOVE The Roth-Steyr Model was developed by George Roth and first issued to the Austro-Hungarian cavalry in 1907. The design included a bolt that extended the full length of the receiver.

Cavalry pistols

The Roth-Steyr Model is an early 20th-century pistol. It was the first self-loader to be adopted by a major power as a service weapon when it entered service with the cavalry arm of the Austro-Hungarian army in 1907. It was a large and heavy pistol that fired a unique rimless 8mm cartridge. The barrel and breech were locked together and are separated during recoil by a cam action that rotated the barrel through 90 degrees to unlock the bolt. Though the pistol had 10-rounds of ammunition, this was not housed in a removable magazine and they had to be loaded into the butt (grip) from the top using a charger (a device to load a

magazine). The pistol had a rather complex safety system: the pistol would reload after it had been fired, but would not recock. The bolt had to be pulled back by hand to reset the mechanism if there was a misfire. This was a deliberate feature as a pistol with a conventional pull would have been too light and prone to accidental discharge if the horse shied or bucked.

ABOVE The 1908 Bergmann-Bayard. German industrialist, Theodor Bergmann, hired a gun designer and developed a series of automatic pistols at the turn of the 20th century. It used its own 9mm Bergmann-Bayard cartridge.

ABOVE The 1935 Browning 9mm GP or Grand Puissance (High-Power) pistol. The pistol had been designed a decade earlier in 1925 by the weapons engineer John Moses Browning.

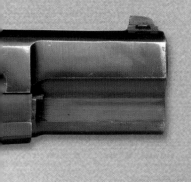

Into the modern age

During the 20th century technological improvements in weapons design meant that they became more reliable, faster and ultimately more deadly. Although revolvers maintained a steady following, the automatic handgun, initially designed by John Moses Browning, became standard issue for police and security forces around the world. Many have become classics: the Luger, Beretta and Browning High-Power remain highly regarded weapons. Two world wars and countless smaller conflicts have only encouraged arms manufacturers to produce lighter, quieter and more powerful weapons. From the Mauser and Tommy gun, via the Sten gun to the Uzi, successive generations have produced ever more efficient weapon systems. In the early years of the 21st century, the wheel seems to have gone full circle. Fail-safe firepower is still critical, but there is now a demand for non-lethal ammunition that will fell an opponent without causing permanent damage.

ABOVE Weapons training for women in 1941. The submachine gun in the foreground is a Sten Mark III and the Sten in the background is a Mark I. The Mark I had a wooden foregrip and forward handle. The Mark III is a simplified version of the Mark I and II.

Guns of the Empire

The weapons of empire used by soldiers and adventurers were often large-calibre revolvers that could be used quickly in a confined space such as a tent or cabin, and long-range rifles used in engagements on the North-West Frontier of India or the open veldt of South Africa. A classic example was the British Webley .455 revolver developed by Webley & Son (Webley & Scott Revolver and Arms Company Ltd from 1897) in the 1870s. A few years later the French Army adopted an 8mm revolver, the Modèle d'Ordonnance (Lebel); it was a revolver that would be taken into the 20th century.

Webley revolvers

The British Webleys were the first "top-break" revolvers with a two-piece frame, which hinged (or broke) at the forward low end for the ejection of cartridge cases and loading. The ejector operated automatically when the frame was broken open and all six empty cases were ejected simultaneously from the cylinder. The cartridges could then be inserted by hand. Designers of revolvers in all calibres adopted the top-break system, as it made for quick reloading – crucial in a short-range firefight. Webleys that had been rechambered for the .45 ACP (a calibre known as .45 Automatic Colt Pistol) round used two three-round half-moon clips (to hold enough ammunition to half-fill the cylinder). This further speeded up reloading.

six-chamber cylinder

hammer

bird's-head hand grip

LEFT The Webley Mark I, adopted by the Royal Navy in 1887, was the first in a series of revolvers produced by this British company.

BELOW The last stand of a British officer at the Battle of Isandlwana in 1879. He has only six rounds in his Webley revolver and little scope to reload quickly.

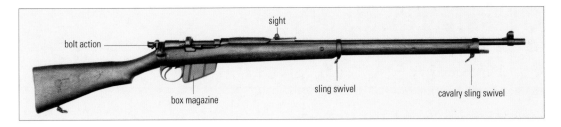

ABOVE The British eight-round .303 Lee Metford, the first repeater rifle. Adopted by the British military in 1888, it encouraged development in small-arms design which, in the case of the Lee Metford, resulted in the introduction of the No. 4 Rifle.

Personal choice

British Army officers and colonial administrators of the Empire would often buy private weapons – handguns that were either superior in design or more convenient to use than standard-issue weapons.

The penalty for this personal choice could be finding ammunition if the weapon was of an unusual calibre. In a remote outpost of the Empire this could literally be a matter of life and death and persuasive words of a salesman in a London gunsmith would ring hollow as the officer looked at his emptying ammunition pouch in a firefight.

In 1887, the British Army and Royal Navy took the first Webley revolver as its official service revolver. This six-shot, double-action revolver was engineered to take the black powder .455 British service cartridge. This cartridge fired a large lead bullet, but because it used black powder it had a relatively slow muzzle velocity of 180m/s (590.55ft/s). Although a smokeless cartridge was then developed, the velocity remained low and could also be fired in early revolvers.

All Webley revolvers were double-action or double-action-only, with a very distinctive barrel shape, a frame lock with lock lever on the left side of the frame, and V-shaped lock spring at the right side.

Trench raids

The generals and political leaders who took Europe to war in August 1914 had not anticipated the possibility of trench warfare, but as the stalemate on the Western Front continued, compact and handy weapons such as the Webley became highly valued by the combatants.

The long bolt-action rifles carried by most infantry in European armies (the exception was the British, with their compact Short Magazine Lee Enfield), were ideal for long-range engagements in open country, but were impractical in the confines of a trench. In the quick, violent fighting patrols known as "trench raids" soldiers carried clubs, knuckle-dusters and knives – silent weapons that could be used in hand-to-hand combat. The revolver or semi-automatic pistol were the only useful firearms in these confined spaces. This was the battleground that would spawn a new weapon: the submachine gun.

The French army revolver

The French 8mm Modèle d'Ordonnance (Lebel) 1892 revolver enjoyed a remarkable longevity in service with the French Army. Adopted in 1892, it was still in service in World War II. The pistol had a conventional swing-out cylinder, with the release button on the right, which made it a user-friendly weapon for left-handed shooters. It had an ingenious system that allowed the left-hand plate to swing forward on a hinge to expose the mechanism for cleaning. Like many weapons developed in the late 19th century the revolver used its own special 8mm ammunition.

LEFT The Webley Mark VI was first produced in 1915 and remained in service until World War II. The revolver, similar in many ways to the Mark V Webley, was issued to British and Commonwealth forces.

Classic revolvers

Revolvers were the classic handguns of the 19th and early 20th centuries. Although they were reliable and robust, reloading under pressure always presented a problem, and the weapon could become clogged by dirt or grit.

The Nagant Model 1895

Widely regarded as a Russian revolver, the Nagant Model 1895G was actually designed in the early 1890s by the Belgian brothers Emile and Leon Nagant. It was first manufactured and used in Russia and adopted by other countries, including Sweden and Poland and, later, the USSR. Local production began in 1898 after the Imperial Russian government had received shipments from Belgium. Technically, the Nagant was almost obsolete when it was adopted in 1895, since revolvers like the American Smith & Wesson Hand Ejectors and Colts with side-opening cylinders were much faster to reload. However, it was not until 1930 that the Nagant Model 1895 finally became obsolete in Russia, enduring as the standard sidearm for more than thirty years. It was widely used and manufactured during World War II, and production finally ceased in the 1950s. The Nagant Model 1895 had some distinctive features, such as the gas-sealed cylinder, which, almost uniquely, made it a revolver suitable for mounting a

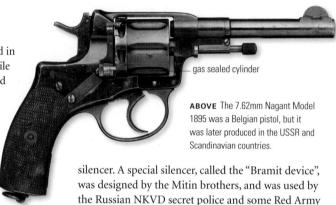

gas sealed cylinder

ABOVE The 7.62mm Nagant Model 1895 was a Belgian pistol, but it was later produced in the USSR and Scandinavian countries.

silencer. A special silencer, called the "Bramit device", was designed by the Mitin brothers, and was used by the Russian NKVD secret police and some Red Army special forces during World War II.

The Webley-Fosbery model 1901

Designed by George Fosbery, a Victorian hero who at the age of 31 had won the Victoria Cross, the Webley-Fosbery is a remarkable weapon. In effect it is an automatic revolver. It opens, loads and unloads like other contemporary Webley revolvers. However, when the pistol is fired, the recoil from the cartridge moves the receiver back in its frame. At this point a cam pin

RIGHT The original caption for this American poster reads: "Cowboy spurring on horse to escape Indians, while shooting it out with a Smith and Wesson pistol." The poster was created in the late 19th century in order to promote Smith & Wesson revolvers.

fixed in the frame engages in zigzag slots in the outer surface of the pistol cylinder, and revolves it half-way toward the next chamber. Simultaneously the pistol's hammer is moved into the cocked position. As the frame-mounted spring returns the receiver forward, the cam pin forces the cylinder to revolve the rest of the way, and the weapon is cocked and ready to fire its next shot. Because of this feature the Webley-Fosbery is fitted with a safety catch – an unusual feature in revolvers.

The weapon was popular with British officers before World War I, who could purchase a revolver made to order. The Webley was available in .455 calibre for the British service cartridge, and later in .38 ACP. With a ready supply of six-round clips, a marksman could attain a rate of fire of 70 rounds per minute (rpm).

The Colt model 1917

With the United States' entry into World War I, manufacturers of the Colt 1911 with the .45 ACP cartridge could not keep up with the demand of the US Army. Two pistols were chosen as a stopgap measure:

ABOVE The Webley-Fosbery .455 Model 1902, a revised version of the 1901 automatic revolver, an ingenious weapon that harnessed recoil to revolve the cylinder.

the Colt New Service pistol and the Smith & Wesson New Century, both of which had been used in action by the British Army on the Western Front. In the light of this combat experience the design of the Smith & Wesson was modified. The Colt New Service was chambered in the .45 ACP cartridge and known as the Colt Model 1917. It could be quickly loaded or unloaded by either three-round stamped metal half-moon clips or six-round full-moon clips.

The "Three Eight"

After World War I the British War Department decided that tank crews and other service personnel who required a sidearm needed a more compact handgun. The result was the Enfield No. 2 Mark I .38 revolver developed at the Royal Small Arms Factory in Enfield between 1926 and 1927. The design was a scaled-down Webley Mark VI with its "break-top" frame and cylinder chambered for six rounds and firing a heavy-grain bullet. The hammer/trigger group was redesigned, with a manual hammer safety lock added, as well as a separate cylinder lock. This revolver was adopted for British military service in 1932 as the Enfield revolver, .38 No. 2 Mark I. After 1938, almost all No. 2 Mark Is were converted into No. 2 Mark I* configuration.

The Enfield No. 2 Mark I* was developed in the late 1930s for the British Tank Corps (part of the Royal Armoured Corps) and was distinguished from the early Mark I by a spurless, double-action-only

ABOVE The early Enfield No. 2 Mark I revolver. This was modified in the late 1930s for the British Army and remained in service until the late 1960s.

hammer, lighter mainspring and reshaped grip side plates. The spurless hammers, which were required by the Royal Armoured Corps, meant that the weapon could be carried in an open holster without snagging on controls or cabling inside the tank.

The Enfield No. 2 Mark I** appeared in 1942 as a simplified, wartime design. These revolvers were similar to the No. 2 Mark I*, but without the hammer stop. After 1945, all the Enfields were recalled and converted into No. 2 Mark I* configuration. Known informally as the "Three Eight" in the British army, the revolver was robust and comfortable to fire.

Enter the semi-automatic

To men who carried revolvers, the semi-automatic pistol initially seemed over-engineered and too complex. It would, however, supersede the revolver in almost all roles in the 20th century. Among its attractions were a more compact shape and ease of reloading.

The John Moses Browning

The first semi-automatic pistol, the FN Browning M1900, was designed by American firearms maker J.M. Browning in *c.*1896, and followed by an improved version in 1897. In 1898, Browning's design was accepted by the Belgian firearms manufacturer Fabrique Nationale de Herstal (FN), who began production of the 1899 model. A year later, in 1900, Belgium adopted the FN Browning M1900, a modified version with a shorter barrel. Use of the Browning Number 1 pistol, as it was also known, spread across Europe as a civilian and police sidearm. The chamber was designed for the 7.6mm x 17mm round, which is also known in the Americas as the .32 AC, a new smokeless round. Manufactured until about 1911, more than 700,000 FN Browning M1900 pistols were made.

An Austrian aristocrat

The 7.63mm Austrian Mannlicher Model 1900 has all the elegance of a duelling pistol from an earlier century. Designed by Ferdinand Ritter von Mannlicher,

hammer   barrel

ABOVE The 7.63mm Austrian Mannlicher Model 1900, designed by Ferdinand Ritter von Mannlicher, has a mechanism which has been likened to that of a fine watch.

ABOVE The FN Browning M1900 was one of the most commercially successful semi-automatic pistols produced before World War I. It was widely used by police forces and bodyguards.

Blowback and recoil operations

Semi-automatic pistols use either blowback, recoil, or other systems to work the mechanisms that fire, expel and reload ammunition:

Blowback-operated weapons use the pressure created from the fired round to push a bolt, located directly behind the round, back and forth against a spring. Its pressure pushes the bolt backwards against the spring and also ejects the spent round from the gun. A new round enters the weapon, and as the compressed spring pushes the bolt forward, the bolt pushes the round into the breech. A pin on the end of the bolt strikes the round and fires it, beginning the cycle again. Blowback weapons are simple and reliable, but they do not form a complete seal at the breech when firing.

Recoil-operated weapons push the barrel and the breech backwards as a unit, along with the bolt. The ejection and reloading cycles are completed during this recoil, and the breech remains sealed during firing. Recoil-operated have more moving parts than blowback-operated weapons. The weapons are heavy but very reliable. As a result, they have relatively low rates of fire.

it is a well-balanced, light pistol with a mechanism that has been likened to that of a fine watch for its precision and action. Von Mannlicher had a good design but knew that it could be improved. The 1900 model was followed by the 1901 with a longer barrel and a repositioned rear sight. In 1903 he produced a model with a box magazine (a device for holding ammunition via a slot on the receiver) in front of the trigger guard and a new mechanism. Although European armies did not take up the 1900 and 1901 models, they became popular in South America. One of the drawbacks of this pistol was that it used a special cartridge made only in Austria, and, to add to the logistic problems, the 1903 pistol had its own specialized ammunition.

The Bergmann-Bayard

This German pistol was one of the first pistols developed to use 9mm calibre ammunition, although this was not Parabellum but a round with a larger

hammer

magazine

LEFT The Bergmann-Bayard (1908 model) had a six- or ten-round detachable magazine but could also be loaded with a stripper clip.

cartridge unique to the Bergmann-Bayard. Produced in 1903, the pistol was adopted by the Danish army in 1905, and later by the Belgian and Spanish armies. It was made at the Bergmann works in Germany and also mass-produced by Peiper of Herstal in Belgium. Although the pistol looks like a Mauser, it has a simpler trigger mechanism and the magazine can be removed. To reload, the magazine can either be filled by hand or, with the action open, a charger can be inserted and the rounds pushed into the magazine – an obvious advantage in a military pistol. Like other weapons of this period it could be fitted with a holster stock that was clipped to the bottom of the butt to convert it into a light carbine (automatic rifle).

The Luger Model 1908

The German Luger self-loading pistol, known in German army service as the Pistole 08 from its year of adoption, was named after George Luger, a designer at the Ludwig Löwe small-arms factory in Berlin. Hugo Borchardt based the Luger's design on an earlier idea, but Luger re-designed the Borchardt locking system (called lock-out) into a much smaller package. The toggle-lock mechanism was complex, but it made the weapon comfortable to fire and therefore more accurate. The first military Lugers were made in 1900 to a Swiss order. The original calibre was 7.65mm, but in 1902 at the request of the German navy, the firm of Deutsche Waffen und Munitionsfabriken (DWM), along with Luger, designed a bigger round. By re-necking the case of the 7.65mm Luger round the 9mm Luger/Parabellum was developed. The standard pistol had an eight-round box magazine and fired the Parabellum round with a maximum effective range of 70m/230ft.

rear sight (flip-up)

toggle link hinge

magazine release

ABOVE The Luger Model 1908, also known as the Pistole 08.

In the 1920s the British firm Vickers manufactured Lugers for export to the Netherlands. The shortest pistol had a 103mm/4.06in barrel, an overall length of 222mm/8.75in and weighed 875g/30.9oz.

The Naval Luger weighed 1,043g/36.8oz, had a 198mm/7.8in barrel, a tangent sight, and could be fitted with a 32-round "snail" magazine, effectively making it a light submachine gun.

Semi-automatic innovation

The German Luger was an elegant design. However, the first .455in self-loading pistol produced in 1904 by the prolific British firm Webley & Scott Revolver and Arms Company Ltd did not have the same appeal, nor was it a commercial success. The American Colt M1911 self-loading pistol, however, proved to be the era's great survivor: with its rugged design and reliability in the front line it remained in service until the Vietnam War.

The .455 Webley & Scott 1910

An improved version of the .455 Webley & Scott pistol introduced in 1910 proved a real success. It was the Royal Navy's standard pistol until the end of World War II and was used by the Royal Air Force, the Metropolitan Police and throughout the Empire. The pistol had a grip safety (a safety mechanism that prevents a gun from being fired, unless the grip is held firmly). On firing the ejection port threw the empty cases upwards and forwards. An ingenious system with the magazine allowed a user to fire hand-loaded single shots with the magazine partially inserted – if he needed a full magazine he simple had to push it completely home and six rounds were immediately available.

bolt   cross key

LEFT The Italian Glisenti Model 1910, sometimes referred to in early records as the Brixia, fired a low-powered 9mm round.

At the same time, the Italian Glisenti M1910 became the first semi-automatic pistol to be adopted by the Italian military. Although a later version was chambered to use the same cartridge case as the 9mm Luger, it was largely replaced by the 9mm Beretta Model 1915 in World War I.

Introduced in 1912, the Austro-Hungarian forces in World War I preferred the Steyr Model 1912 pistol over the range of other standard issue pistols. It was also used by the Romanians and Germans. Officially named "Selbstiade Pistol M12" by the Austrian Army, the pistol was also informally known as the Steyr-Hahn ("Steyr with a hammer") to distinguish it from the Roth-Steyr. During World War II, a number of weapons were re-barrelled for use by the German Army to take the 9mm Parabellum cartridge. About 250,000 of the pistols were made before production ceased in 1919.

The .45 Colt 1911

The giant in the world of self-loading or semi-automatic pistols is the .45 Colt 1911, a US government model developed from a Browning design. It is a much more reliable pistol under muddy or sandy conditions than weapons like the Luger, because it does not have so much of its working mechanism exposed. The United States adopted this

ABOVE A British sergeant armed with a Colt .45 M1911A1 semi-automatic pistol stalks a German sniper in Italy, 1943. The American-made Colt was widely issued to Allied forces and was a popular sidearm. Its .45 rounds made it highly effective at short range.

pistol in 1911 and modified it in 1921. It was redesignated the M1911A1. American servicemen carried the .45 Colt through World War II, the wars in Korea and Vietnam, and it was also widely issued to Allied forces. It was 215mm/8.46in long, weighed 1,105g/39oz, had a seven-round magazine, and a muzzle velocity of 262m/s (860ft/s). Ammunition from the Colt 1911A1 was compatible with the Thompson submachine gun Model 1928A1.

ABOVE The Austrian Steyr Model 1912 was converted by the Germans in World War II to take 9mm Parabellum.

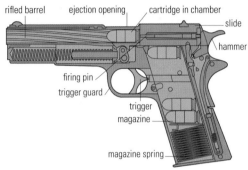

ABOVE A cross-section of a .45 Colt pistol. The rounds are held under compression in the magazine and each time the slide moves to the rear – either on firing or when the pistol is cocked, a round is fed into the chamber. Pressing the trigger releases the mechanism that allows the spring-loaded hammer to hit the firing pin.

Soviet TT pistols

About this time in the Soviet Union, Fedor V. Tokarev began developing a series of pistols with the prefix TT for Tula-Tokarev, or Tulskiy Tokarev (meaning Tokarev from Tula). The 7.62mm TT33 was developed in 1930 and adopted by the Soviet Army the same year. In 1933, the pistol was improved with a new locking system and a different disconnector (used to prevent the gun from firing unless a magazine is inserted). It looked rather like the 1903 Colt .38 Pocket Automatic Pistol and could be disassembled (field stripped) in a similar way to the Colt 1911.

The pistol was single-action, and recoil-operated in the same way as the Browning design, with an eight-round single-stack magazine.

Japanese Nambu Type 14

Across the Pacific, the 14th Year Pistol designed by Japanese General Kijiro Nambu was an improved and cheaper version of an earlier, similar Japanese pistol. Externally it looked like a Luger, but the locking system was based on a different system. The Imperial Japanese Army adopted it in 1925 as the official sidearm with the designation Type 14 Pistol. One unusual version, nicknamed "Kiska", had a very large trigger guard designed for use by a soldier wearing thick gloves in the bitter winters of northern Japan, mainland China and Korea.

The Nambu Type 14 was maligned as a very poorly designed weapon. It fired an underpowered 8mm cartridge with a muzzle velocity of 290m/s (950ft/s). It was difficult to reach and operate the safety catch, and if the magazine release catch became clogged it

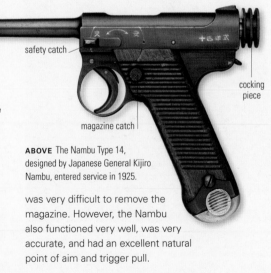

ABOVE The Nambu Type 14, designed by Japanese General Kijiro Nambu, entered service in 1925.

was very difficult to remove the magazine. However, the Nambu also functioned very well, was very accurate, and had an excellent natural point of aim and trigger pull.

The submachine gun

The need for a submachine gun had its origins in World War I, as a method of clearing the enemy trenches of soldiers at close-range. Fast-firing semi-automatic pistols fitted with a detachable stock pointed the way to the introduction of the first submachine gun (SMG). By the close of that war the Germans and Italians had produced the first submachine guns – the German weapon being a classic design.

twin barrels

rear grips

bipod mount

ABOVE The Italian 9mm Glisenti Twin Villar Perosa. Though this unusual-looking weapon is a world away from modern submachine guns, it was in fact the world's first submachine gun. The Italians failed to see its potential at first and used it as a static light machine gun.

The Mauser pistol

Known as the "Military Model", the German 1895 Broomhandle Mauser pistol was never an official issue weapon, although it was popular with officers as a private-purchase sidearm. Although it had been made in 7.63mm with a special 9mm Mauser cartridge, by 1914 many had been converted to 9mm Parabellum (the most standard calibre). The standard model had a built-in ten-round magazine, fed by a clip from the top. A fully automatic type was also produced in limited numbers and used by German troops as a light machine pistol – a submachine gun by any other name.

ABOVE A German Waffen-SS soldier takes aim with a private-purchase Mauser military pistol, converted from the original 7.63mm model to 9mm Parabellum. Fitted with a shoulder stock, this weapon could be fired very accurately by experienced shots, particularly against static targets.

The first submachine guns

In 1915 Italy became the first country to adopt a submachine gun, known as either the "Villar Perosa", or the "Fiat" depending on its place of manufacture, or the "Revelli" after the designer Abiel Revelli. The original production had no stock, and two barrels with two 25-round box magazines and thumb-type triggers. Like all other early Italian submachine guns, it was chambered for 9mm Glisenti ammunition, basically a low-powered 9mm Parabellum cartridge. Although it was portable, it was largely used as a static machine gun. By 1918 Italian soldiers were re-equipped with a Beretta modification of the design and began to use the submachine gun in an assault role.

The German Bergmann MaschinenPistole 18 or MP18 was designed by Louis Schmeisser in 1917 and incorporated many of the features that would become standard in early submachine guns. The German army ordered 50,000 and around 10,000 had been delivered in time for the *Kaiserschlacht*, the 1918 offensive on the Western Front. After the war, stocks of the gun were taken over by the French Army which substituted a 20- or 32-round box magazine that loaded from the left. This feed mechanism was adopted when production recommenced in 1928 under licence in Belgium.

At the close of World War I the Beretta M1918 (or strictly the Beretta Moschetto Automatico 1918) was designed by the talented young engineer Tullio

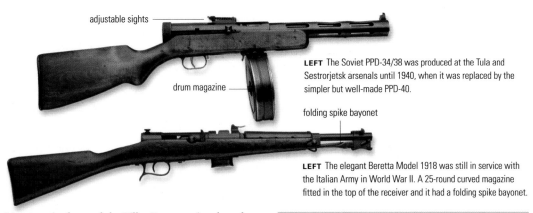

adjustable sights

drum magazine

LEFT The Soviet PPD-34/38 was produced at the Tula and Sestrorjetsk arsenals until 1940, when it was replaced by the simpler but well-made PPD-40.

folding spike bayonet

LEFT The elegant Beretta Model 1918 was still in service with the Italian Army in World War II. A 25-round curved magazine fitted in the top of the receiver and it had a folding spike bayonet.

Marengoni, who used the Villar Perosa action, barrel, body and feed system for the cartridges. A new trigger and stock were added, with an ejection chute under the ejection slot. Finally, the bayonet catch and folding bayonet, two components from the Carcano military carbine, were added. It was issued to *Arditi* (assault troops) in the Italian army.

The PPD (Pistolet-Pulemyot Degtyarova) was developed by the Russian small arms designer Vassili A. Degtyarov in 1934 and adopted by the Red Army a year later. With a 25-round box magazine, it went into limited production with the designation PPD-34. Following modifications in 1938, it was produced until 1939 as the PPD-34/38 with a newly developed 71-round drum magazine. Although a new version was developed in 1940 with a two-part stock designed to accept a new pattern of 71-round drum magazine, by June 1941 it became clear that the PPD-40 was not suited to wartime mass production, and it was replaced by the more efficient and inexpensive PPSh-41.

The French 7.65mm Mitraillette MAS 38 submachine gun introduced in 1938 had evolved from the experimental MAS 35 and was an accurate short-range weapon machined from solid metal. It fired from a 32-round box magazine with a cyclic rate of between 600 and 700 rpm. Following the fall of France in 1940 the Germans kept production going at Manufacture d'Armes de St Etienne for their own use until 1944.

Tullio Marengoni and Pietro Beretta

Firearms historians have described Tullio Marengoni as Europe's John Browning since his designs that began with a hammerless shotgun in the early 1900s continued to dominate every significant Beretta weapon for the next sixty years. Beretta and Marengoni had a unique working relationship, and would often work together all through the night. Marengoni came up with the concepts and Beretta engineered the working weapons from his designs. Their wide-ranging output included the monobloc breech for shotgun barrels, the classic Beretta over/under shotgun line, and the pistol design that would evolve into the official US military sidearm, the modern Model 92/M9. Marengoni remained a consultant for Beretta for several years after his official retirement in 1956.

BELOW The early 32-round "snail" magazine designed for the 9mm German MP18 submachine gun proved complex and prone to stoppages.

barrel jacket   magazine housing

RIGHT The MP18 would have been issued on a scale of six guns per company, with gunners supported by a second soldier carrying extra ammunition. However, though World War I ended before this could be fully tested, it was a pointer to the future.

The trench broom

The American M1928, the Thompson Model 1928 "Trench Broom" or "Tommy Gun", was the brainchild of John T. Thompson, who helped develop the M1903 Springfield rifle and the .45 M1911 pistol (shortly after his retirement from the army in 1918). He began work on a handy, fast-firing weapon for use in attacks through field fortifications.

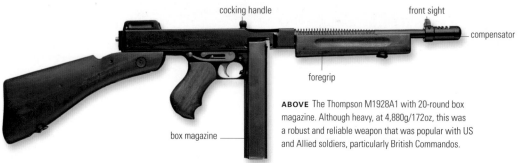

cocking handle   front sight   compensator   foregrip   box magazine

ABOVE The Thompson M1928A1 with 20-round box magazine. Although heavy, at 4,880g/172oz, this was a robust and reliable weapon that was popular with US and Allied soldiers, particularly British Commandos.

The Thompson submachine gun

Drawing on experience of fighting on the Western Front, John T. Thompson recognized that the .45 bullet used in the M1911 pistol would transform a fully automatic weapon. Early in 1920, a prototype capable of firing 800 rpm was produced by Thompson's company, Auto-Ordnance.

Despite its performance in trials, neither the US Army nor the Marine Corps adopted the Thompson. Nevertheless, Thompson was able to sign a contract with Colt for 15,000 "Thompson Submachine Gun, Model of 1921".

ABOVE A French soldier, armed and equipped by the US Army, on guard in 1945 with his Thompson M1A1. Weighing 4,740g/167.2oz, this was the final simplified version of the submachine gun, with a fixed firing pin.

There were no further orders for submachine guns until the eve of the United States' entry into World War II. From 1942, the orders were huge. The reliability of the Thompson to outperform other submachine guns in adverse conditions such as dirt, mud and rain was one of its main assets. British Commandos retained their Thompsons after the Sten had become more widely available, and went ashore at D-Day in 1944 armed with them. The main drawbacks of the gun were weight, inaccuracy at ranges over 45m/50yd, and lack of penetrating power. However, as a close-range weapon it was devastating.

The "Tommy Gun"

In 1940 the British Government rushed to purchase the Model 1928A1, Thompson Machine Carbine (TMC) or "Tommy Gun". Senior staff officers in the British army had referred to submachine guns disparagingly as "gangster weapons" because of their gangland reputation in Prohibition America, but the reservations of the British War Office were forgotten in the face of the threat of German invasion, and orders were placed for the Model 1928.

This gun saw service in North Africa and Italy. It weighed 4,800g/172oz was 857mm/33.75in long, had sights set out to 548m/1,800ft, and had a cyclic rate of 600 to 725 rpm with a 20- or 30-round box or a 50-round drum magazine. Most soldiers favoured the 20-round magazine that fitted into a pouch and made for easier movement in close country.

In 1942 the original Thompson M1928 had the cocking handle (which compresses the spring behind the bolt) on top of the receiver and utilized the Blish Lock system of operation. The Blish Lock system worked on a principle of static friction in a delayed-blowback breech lock. The gun was redesigned as the Thompson M1A1. The M1A1 used a simple blowback system of operation and the cocking handle was moved to the side. To cut manufacturing costs, the Cutts compensator (a device which alleviates the tendency of the muzzle to rise when firing on full automatic), the finned barrel, the fingered fore-grip, and the flip-up adjustable rear-sight were either simplified or eliminated. The M1A1 did not accept the drum magazine, which was abandoned for reasons including the fact that it was bulky and that the rounds rattled, a noise that could be fatal on night patrols. By the end of World War II, more than one million M1 and M1A1 submachine guns had been made.

RIGHT A sales brochure for the Thompson submachine gun, published by the Auto-Ordnance Corporation. The "Tommy Gun" was used by police and gangsters in the 1920s and 1930s. This led the British War Office to describe them as gangster weapons and reject them in the 1930s.

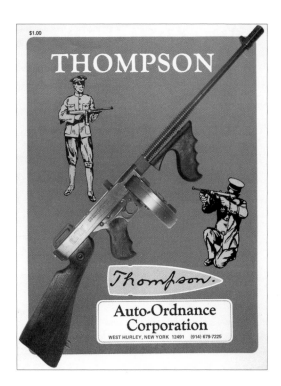

$1.00

THOMPSON

Thompson

Auto-Ordnance Corporation
WEST HURLEY, NEW YORK 12491   (914) 679-7225

Gangster weapons

Compared with the thousands of Thompson submachine guns used properly by law-abiding Americans in the course of their duty, only a very small number of Thompsons were misused by Depression and Prohibition-era gangsters in the United States. Despite this, the weapon got a bad name in the late 1920s and 1930s.

The best-known front-page crime remains the notorious "St. Valentine's Day Massacre" of 1929 when henchmen of the gang-leader Al Capone, who had conveniently left town for Florida, shot down seven members of the "Bugs" Moran gang in a Chicago garage.

On a summer night in 1933, Oklahoma City oilman Charles Urschel was kidnapped from his home by gangster George Kelly. Kelly was armed with a Thompson submachine gun and this crime earned him the nickname "Machine Gun" Kelly.

"Public Enemy No.1" John Dillinger stole Thompsons from small-town police stations. Both the Dillinger–Nelson gang and Charles Arthur "Pretty Boy" Floyd customized their Thompsons, making them easier to conceal and fire one-handed. This was done by removing the butt and replacing the 50-round drum magazine with a 20-round box magazine.

Hollywood also played its part in giving the gun a bad reputation, with Jimmy Cagney and Edward G. Robinson playing Thompson-toting hard men in violent movies like *The Public Enemy* and *Little Caesar*.

LEFT John Dillinger, "Public Enemy No. 1", one of the American Prohibition-era gangsters who used the Thompson M1928 submachine gun with a 20-round box magazine.

The 1930s: rearmament

When Adolf Hitler came to power in Germany in 1933, he introduced a vast programme of rearmament, in contravention of the Versailles Treaty of 1919. Some superb designs and concepts were developed in the late 1930s and early 1940s that have long outlasted the life of the Third Reich. The 9mm Walther Pistole 38 or P38 was designed to replace the P08 Luger and the compact PP and PPK pistols were introduced. In Italy Beretta produced the elegant Pistol Automatica Beretta Modello 1934/38 and the 8mm Mitra Beretta MAB38 submachine guns. However, the Browning 9mm High-Power pistol developed in the 1930s would be one of the most successful pistols of the period.

The Walther pistole 38 (P38)

The German Walther P38 was safe, easy, and cheap to manufacture.The P38 has the distinctive feature of a positive safety catch that prevents accidental firing, even when cocked with a round in the breech. A pin indicator, which could be felt in the dark, showed if a round was loaded. The magazine held eight rounds that could be fired in 20 seconds with an effective range of 50m/164ft. Between 1939 and 1945 over one million P38s were made and it is widely acknowledged to have been the best military pistol of World War II.

ABOVE The German wartime Walther P38 would return after 1945 as the lightweight Bundeswehr Model P1 pistol.

The Walther PPK

In 1928, before the Nazis came to power, Walther produced the Model PP or Polizei Pistole, a compact weapon with an eight-round magazine designed for police use. The PPK or Polizei Pistole Kriminal was only 150mm/5.91in long compared to the PP's 170mm/6.7in, and was intended for police working in plain clothes. During World War II it was the personal weapon of choice for senior officers in the German armed forces, and it remained in service after the war, later becoming famous as the weapon of Agent 007, Ian Fleming's fictional spy James Bond.

Italian designs

In Italy another small-arms manufacturer with a worldwide reputation was producing new designs. The Pistol Automatica Beretta Modello 1934/38 was a handy blowback-operated weapon that weighed 616g/21.76oz. The finish was excellent and it became one of the most prized war trophies for the Allies.

The 8mm Mitra Beretta MAB38 submachine gun introduced in 1938 was based on the Villar Perosa. It had two triggers, one for semi-automatic and one for automatic fire. It fired from 10-, 20- or 40-round magazines that could be loaded by hand or by specific

ABOVE A British sergeant proudly displays the FN Browning 9mm High-Power pistol, captured during fighting in north-west Europe. The Brownings were popular with Allied and German soldiers because of their capacious double-stacked 13-round magazine.

adjustable rear sights

tools. The maximum range for the pistol-sized bullet (the same as the Beretta 34 pistol) was allegedly 1,000m/ 3,281ft; in reality the effective range was around 100m/328ft.

The Browning high-power

In 1935 the Belgian firm of Fabrique National (FN) began production of the Browning 9mm GP or Grand Puissance (High-Power) pistol. They offered two versions: the standard with fixed sights and the "Adjustable Rearsight Model" with tangent rear sights, graduated up to 500m/1,640ft and a slot in the pistol grip to take a detachable wooden stock. The pistol had been designed a decade earlier by the weapons engineer John Moses Browning, and granted a US patent in February 1927 three months after his death. This superb weapon, with its double-stacked 13-round magazine, would be produced in Canada by Inglis for use by Allied troops,

ABOVE The 1935 FN Browning High-Power pistol had adjustable tangent rear sights. A stock could be fitted to the grip which, in theory, would enable the pistol to be used against targets at 500m/1,640ft.

while the FN weapons were eagerly taken up by the Germans following the occupation of Belgium in 1940. Prior to 1939 the pistol had already been adopted as a service sidearm in Belgium, Denmark, Holland, Lithuania and Romania. In the late 20th century it remained in service in more than 55 countries and the improved 9mm Browning High-Power Mark 3 was still in production at FN.

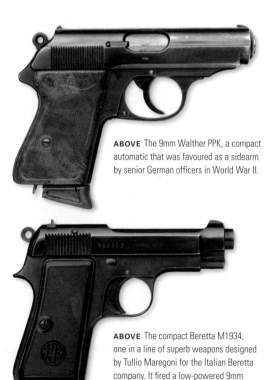

ABOVE The 9mm Walther PPK, a compact automatic that was favoured as a sidearm by senior German officers in World War II.

ABOVE The compact Beretta M1934, one in a line of superb weapons designed by Tullio Maregoni for the Italian Beretta company. It fired a low-powered 9mm Short or .38 Auto round.

9mm Parabellum

This is a world standard for pistol and submachine gun ammunition used and manufactured throughout the world. The 9mm Parabellum cartridge was developed in 1902 by George Luger. "Parabellum" was taken from the ancient Latin saying *Si vis pacem, para bellum*, meaning "If you want peace, prepare for war".

The original cartridge used an 8g full metal or a semi-jacketed truncated-cone bullet. The jackets were nickel-plated steel, although later copper was used. The charge (0.35g) of smokeless powder gave a muzzle velocity of 310m/s (1,017ft/s).

The name "9mm Luger" was never an official designation but the Luger name was registered in 1923 as a marketing ploy by A.F. Stoeger, the American company who imported the ammunition in the years between the wars.

RIGHT The term Parabellum has been used in the naming of a number of cartridges. However, a Parabellum is most likely to refer to a 9mm cartridge.

German submachine guns

The origins of the submachine gun in German service during World War II date back to 1916 when the Imperial German Army tasked the small-arms designer Hugo Schmeisser to produce a short-range rapid-fire weapon. It was ready for the 1918 offensive and was designated the Machine Pistol 18 or MP18. The first German submachine gun, the 9mm MP34/35, was designed by Bergmann. However, it was the MP38 and MP40 (the "Schmeisser") that would be seen as the classic German submachine gun of World War II.

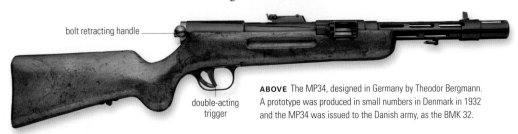

bolt retracting handle

double-acting trigger

ABOVE The MP34, designed in Germany by Theodor Bergmann. A prototype was produced in small numbers in Denmark in 1932 and the MP34 was issued to the Danish army, as the BMK 32.

ABOVE A dramatic propaganda poster featuring a highly decorated German officer leading an assault armed with an MP38.

The MP34/35

Production of the MP34/35 began in Germany in 1934. The bulk of the stock made by Junker & Ruh AG at Karlsruhe went to the Waffen-SS, the Nazi's fighting force. The trigger was unusual: a partial pull produced single shots, and automatic followed if it was pulled fully to the rear. Its most distinctive feature was that the magazine loaded from the right and, when fired on automatic, had a fearsome cyclic rate of 650 rpm.

The Erma

The 9mm MP Erma (or EMP) was developed from an earlier Heinrich Vollmer design from the 1920s and was designed by Heinrich Vollmer and Berthold Giepel, both co-directors of Erma Werke. It remained in front line service until 1942. A silenced variant of the EMP was used to equip the Milice (the French security police who operated under German control) between 1940 and 1944. It had a lower muzzle velocity and rate of fire. The Erma submachine gun has unfairly become known by the name of "Schmeisser" after the senior designer at Haenel of Suhl. Haenel manufactured the weapon, and perhaps that is why Hugo Schmeisser has received the credit for the MP38, 38/40 and 40.

It was the first submachine gun to have a folding metal butt, which reduced its size from 833mm/32.8in to 630mm/24.8in and made it ideal for paratroops and vehicle crews. It had a distinctive lug called a "resting

tunnel foresight

resting bar

folding butt

ABOVE The MP38, first produced by the Erma Werke at Erfurt, included a folding metal butt and was also unusual in its use of plastic.

plastic grip

bar" below the muzzle so that it could be fired on automatic through weapons ports or over the side of vehicles with no danger of the vibration causing it to slip back inside. From the resting bar a metal fin called a "cooling strip" ran back to the receiver. The pistol grip and trigger combination drew on the US Thompson M1928 for inspiration – earlier submachine guns retained a carbine-style wooden stock. It was the first submachine gun to use plastic in the form of Bakelite in its construction.

The MP38 went into production, and in the campaign in Poland in 1939 it soon emerged that the weapon had a dangerous fault. When the submachine gun was cocked the bolt could easily be knocked forward, accidentally causing it to fire. An improvised solution was a leather collar that fitted over the barrel with a strap that held the cocking handle. At the factory a simple safety catch was produced and these modified weapons were designated MP38/40.

Cost cutting and fire power

The drive to cut production costs and speed manufacture led to the MP40. This weapon had many of the external features of the MP38/40. In the new weapon, machining was reduced to a minimum and steel pressings and welds used wherever possible. In Russia, German soldiers armed with the MP40 found themselves out-gunned by the Soviet PPSh-41 submachine gun with its 71-round magazine. To address this problem, late in 1943 Erma introduced the MP40/1. This consisted of a special housing which took two 30-round magazines fitted side by side. While this effectively produced a 60-round weapon, it also increased the weight to 5,402g/190.56oz.

By the end of World War II, some 1,047,000 MP40s had been manufactured. It is reported that an MP40 was the weapon that was used to kill the Italian fascist leader Benito Mussolini when Communist partisans captured him in 1945. After the war it was used by the French, and remained in service as the Maskin 9mm M40 with the Norwegian Army into the 1980s.

"The Schmeisser"

The MP38 and MP40 submachine guns, known loosely by the Allies in World War II as the "Schmeisser", were issued to German Non-Commissioned Officers (NCOs) commanding MG34 and MG42 machine-gun crews. This ensured that if the gun crew was suddenly attacked by an enemy they had a weapon that could produce a high volume of short-range fire. Powerful as the machine gun was it could be difficult to handle in a confined space, unlike the handy submachine gun. The NCO was also issued with binoculars that allowed him to locate distant targets for the machine-gun crew.

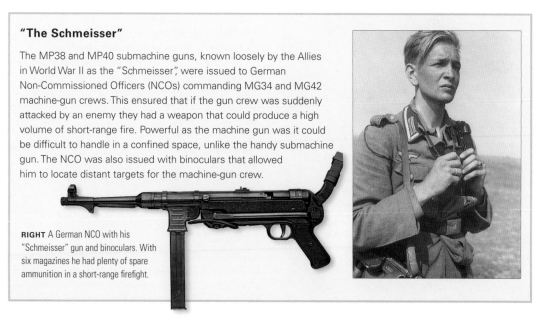

RIGHT A German NCO with his "Schmeisser" gun and binoculars. With six magazines he had plenty of spare ammunition in a short-range firefight.

Fast and cheap

During World War II some weapons were developed at short notice, while others were refined for faster, cheaper production. The definitive examples are the British Sten and the Soviet PPSh-41. The distinctive Owen gun was produced in Australia, and would serve from World War II until Vietnam, while the US developed the UD M42 or Marlin gun and the highly successful M3 "grease gun".

The Owen gun

By 1939 in Australia, Evelyn Owen had developed his first automatic weapon, chambered for the .22 Long Rifle (LR) cartridge, and offered it to the Australian army, which rejected the curious-looking weapon. However, Owen persisted and by 1941 the Lysaghts Newcastle Works in New

sights

retractable butt

barrel

box magazine

ABOVE The American .45 M3 or "grease gun", introduced in 1943, incorporated several innovative design features.

South Wales had produced several more prototypes in different calibres, including .45 ACP, 9mm Parabellum and even .38 Special revolver cartridges. The 9mm prototype weapon was tested successfully against the Thompson and Sten submachine guns. The Owen gun was adopted in 1942 and manufactured until 1945 in three basic versions: Mark 1-42, Mark 1-43 (or Mark 1 Wood Butt), and Mark 2. About 45,000 Owen submachine guns were made by Lysaghts, and these remained in service with Australian forces until the 1960s, through World War II, Korea and Vietnam. Although rather heavy, the robust, reliable and simple Owen gun was well liked by Australian soldiers.

The Marlin gun

The American 9mm submachine gun that would be widely known as the Marlin gun, but had the official designation United Defense M42, was designed between 1941 and 1942 by Carl G. Swebilius, a small-arms designer working for the High Standard

LEFT An Australian soldier armed with an Owen Mark 1. The weapon was well-made and robust and the bolt was protected against mud and dirt. The vertical 33-round magazine ensured a reliable feed, combining gravity and a spring, that compared favourably with the Thompson.

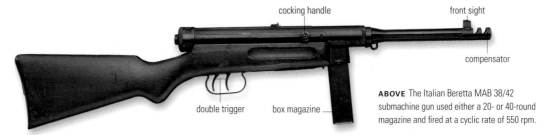

cocking handle

front sight

compensator

double trigger   box magazine

ABOVE The Italian Beretta MAB 38/42 submachine gun used either a 20- or 40-round magazine and fired at a cyclic rate of 550 rpm.

Manufacturing Company. The first UD M42 submachine guns were made by High Standard and Marlin to a Dutch order, and most were shipped to the Dutch East Indies just prior to the Japanese invasion. The remaining 15,000 guns made by the company were purchased by the US government and mostly used by Office of Strategic Services (OSS) operatives and supplied to Resistance groups, notably in Crete. The UD M42 was a somewhat complicated but well-made weapon that fired from a 20-round magazine. In an unusual development, twin reversed magazines were produced for quick reloading.

The M3 "grease gun"

The American M3, a submachine gun produced in 1943 as a low-cost substitute for the Thompson, with several updated design features. Accessories were included, such as an oiler in the weapon's pistol grip and the ingenious wire stock that could be used as a screwdriver, a spanner to unscrew the barrel, a cleaning rod and a magazine filling tool. Though the M3 only fired on full automatic, the cyclic rate was notably low at 450 rounds per minute and the straight line of recoil thrust made it easier to control. An experienced soldier could fire single shots. Before firing, the ejection port cover had to be opened manually by the operator; this also functioned as the weapon's safety catch.

The exterior of the weapon was formed from two pressed-metal shells, welded together, while the barrel of the M3 was secured by a simple nut, and the bolt rode on two guide rods inside the receiver.

This provided clearance between the bolt and receiver, preventing dirt from jamming the weapon, and making it a "soldier friendly" weapon in combat. Its main drawback was that the magazine had a tendency to jam or fall out of the housing if the magazine catch was accidentally depressed. Its unusual appearance earned it the nickname the "grease gun".

The M3 had a 30-round single position feed box magazine and fired the .45 ACP cartridge used in the M1911A1 pistol and Thompson submachine gun. It remained in service with the US Army in Korea, Vietnam and even with some units in the First Gulf War of 1990–91.

MAB38

The Italian MAB38, known as the Moschetto Automatico, was manufactured in the Beretta factory at Brescia. Until 1943, it was only available to parachute units, Carabinieri and Polizia Africa Italiana. It became more widespread in the Armed Forces of the Republic of Salò, partisans and Italian units fighting under Allied command. A later version, the MAB 38/42, was lighter, shorter and had a higher rate of fire. It remained in service well into the 1970s with the Carabinieri and police.

BELOW The UD M42 submachine gun, also known as the Marlin Gun, had a double 20-round magazine and fired at a cyclic rate similar to the Thompson M1 and M1A1 of 700 rpm.

grip

double 20-round magazine

ejection bolt

The Sten gun

Following the fall of France and the evacuation at Dunkirk in 1940, the British faced the prospect of imminent invasion by a well-armed German Army, and urgently needed an inexpensive automatic weapon. The answer was a British submachine gun designed by Major R.V. Shepherd and H.J. Turpin at the Royal Small Arms Factory at Enfield. It was named the "S.T.EN" as a tribute to Shepherd's and Turpin's ingenuity and to the factory.

Powerfully simple

The Sten went into mass production and, by the end of the war, an estimated 3.25 million had been made in Britain and Canada. The guns were cheap and easy to make. The Sten weighed about 3,500g/123oz and all marks of the weapon were blowback operated, with a cyclic rate of 540 rpm and a muzzle velocity of 366m/s (1,201ft/s). Magazine capacity was 32 rounds, but since there were always feed problems this was normally kept down to 30. Between 1940 and 1945 the Sten was modified and improved, going through numerous marks.

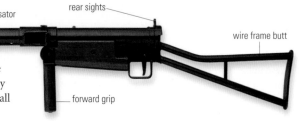

flash hider and compensator · rear sights · wire frame butt · forward grip

ABOVE The Sten Mark II entered service in the summer of 1941, but was not well received by soldiers on account of its rough and crudely made appearance.

The development of the Sten

The Sten Mark I, which featured a flash hider (a device to reduce muzzle flash on firing), wooden furniture, and folding hand grip, was quickly replaced by the Sten Mark II, which saw widespread issue. The fact that the Mark II was compact and easy to take apart made it easy to hide, and it was a weapon of choice for many European Resistance groups. It was also very useful that the similarity of the magazine to the German MP40 meant that it could use captured German 9mm ammunition. However, as with the German MP40, jamming could sometimes be a problem. Easy to manufacture, it was a simply built weapon composed of 47 steel parts that were welded, riveted or pushed together. The bolt and barrel were the only machine-made components. Two million Sten Mark II weapons were manufactured.

One of the requirements in the original design of the Sten was that it could be made by manufacturers who were not specialist gunsmiths. This was exemplified by the Mark III, with its fixed barrel and all-in-one body and casing welded shut along the top. It was built by a firm of toy makers, Lines Brothers. The Sten Mark III was issued to troops in time for the Normandy landings in 1944. Surprisingly, although this was possibly the best Sten version of all, it was not produced in large numbers.

ABOVE Home Guard weapons training for women in 1941. The gun in the foreground is a Sten Mark III and the other is a Mark I. The weapons are not cocked as the women are practising the basic techniques of holding and aiming.

cocking handle

silencer

wire frame butt

fabric grip

ABOVE The Sten Mark II (S) was introduced in 1943. It was popular with the Resistance and was designed to be fired on single shot. Although it was possible to fire it on automatic, the silencer was burned out when bullets were fired at 450 rpm.

Although the Sten was issued to vehicle crews, despatch riders, and those who had no need for a long-range weapon, it was principally issued to infantry battalions, especially platoon commanders, platoon sergeants, and section commanders. Although senior officers could carry a pistol as a personal weapon, many preferred the extra firepower of a Sten.

The Sten was not popular among troops because, besides jamming, it could be fired accidentally if the cocking handle snagged on clothing, or if it was dropped. This happened to Corporal Proctor of the Somerset Light Infantry during operations in Holland in 1944. A soldier slipped, dropped his Sten, and the weapon hit the ground and fired a single shot that hit Proctor in the groin. The platoon commander, Lt. Jary, condemned the weapon as "designed with little but cheap mass production in mind…. It had no locking device at the moment of firing and consequently, if the weapon was jolted sufficiently for the bolt to slide back and engage the stop round in the magazine, this would be fed into the breech and fired."

Stens in action

The Sten first saw action at the disastrous raid at Dieppe in August 1942. In the weeks prior to the raid, Canadian soldiers had to constantly adjust and test their Stens to make sure they worked properly. The first raid, Operation *Jubilee*, was cancelled in July and the Stens were withdrawn. A day before the remounting of the raid, now called Operation *Rutter*, brand new Stens, still in their crates and packed in grease, were issued to some unimpressed soldiers.

German copies

Despite these deficiencies, its ease of manufacture and low cost encouraged the Germans to copy it in 1944–45, producing weapons designated variously as the 9mm MaschinenPistole 3008, VolksmaschinenPistole, Gerät Potsdam and MaschinenPistole 749 (e). Having been so widely produced, the Sten had a lethal afterlife post-1945. The US Army technical manual *Unconventional Warfare Devices and Techniques* includes an example of a copy of the Mark II captured from the Viet Cong in 1964 during the war in Vietnam.

ABOVE Captain Brian Priday, second-in-command of 2nd Batallion of the Oxford and Bucks Light Infantry, armed with a Sten Mark V with fixed bayonet, Normandy 6 June 1944. Two key bridges had been captured across the river Orne and Orne Canal.

Soviet firepower

The weapon that would influence Soviet submachine-gun design and so eventually the war on the Eastern Front was the Suomi ("Finland") submachine gun. Used by the Finnish army in 1931 as the Suomi-KP Model 1931, it was produced by Finnish arms-designer Aimo Lachti between 1920–30. It would shape the design of the hugely successful Soviet PPSh-41 submachine gun.

bolt handle

drum magazine
(various sizes used)

ABOVE The Finnish Suomi-KP M/31 submachine gun was an exceptionally well made weapon. It weighed 4.676kg/10.3lb because many of its components were made from solid metal. As a result it was very robust and would remain in service for many years.

The Suomi

The Finnish Suomi-KP Model 1931 was also known simply as the KP-31, KP – *Konepistooli* or Automatic Pistol – and was manufactured in Finland by Tikkakoski Oy, licensed to Denmark, Sweden and Switzerland, and also widely exported. The Suomi was used very effectively during the Winter War of 1939–40 when the Soviet Union attacked Finland, and the experience of coming under fire from submachine guns with 71-round magazines had a profound impact on Soviet forces. Manufacture of the Suomi ceased in Finland in 1944, but it was used well into the 1990s when it was replaced in the Finnish army by assault rifles.

The PPSh-41

Following the German invasion of the Soviet Union in 1941 when the Russians lost huge quantities of small arms and much of their engineering capability, there was an urgent demand for a light and simple weapon capable of a high volume of fire. The answer to this was the robust and very effective Pistolet-Pulemet Shpagina 1941g or PPSh-41 7.62mm submachine gun, designed by Georgii Shpagin. It was much cheaper and quicker to make than earlier Soviet submachine guns because there were no screws or bolts on it, and all metal parts were stamped or brazed. It weighed 3,500g/123oz and had a 71-round drum magazine based on the Suomi submachine gun, or a 35-round box magazine. It fired at 900 rpm, a rate of fire that in Korea would earn it the nickname the "burp gun".

During World War II Soviet soldiers knew their submachine gun affectionately as the *Pah-pah-shah*, or *Shpagin*. The gun used barrels taken from bolt-action Mosin Nagant M1891/30 rifles that were chromed to reduce corrosion and wear. Stripping was simple, as the receiver hinged open to reveal the bolt and spring.

LEFT A Finnish soldier in a trench on the Eastern Front, holding a Suomi with a 20-round magazine. The gun had a 318mm/12.5in barrel, which made it very accurate over longer ranges.

There was no selector lever (for single shots or automatic) on some of the late models, when the gun was capable of only automatic fire, and though this was high, a rudimentary compensator and muzzle brake helped to reduce muzzle climb (the upward movement of a gun as a result of recoil). The weapon had a few drawbacks: if dropped, it could accidentally discharge, reloading was difficult and the drums were prone to jamming, something which didn't occur with the box magazines.

About five million PPSh-41s had been made by 1945, and the Soviets adapted their infantry tactics to take full advantage of such huge numbers, often equipping complete units, notably "tank descent" infantry, with the submachine gun. The standard equipment load for a soldier seems to have been the one drum and five- or six-box magazines. Before magazines were introduced, it appears that soldiers would have been equipped with two drums. The short range of these weapons meant that these units had to close with their enemy in something of the style of a bayonet charge – a tactic that terrified the enemy on the Eastern Front. The Finns captured a little over 4,000 Soviet 7.62mm submachine guns during World War II. This was too small a number to justify retooling to

sights

angled barrel jacket acts as compensator

71-round drum magazine

ABOVE The 7.62mm PPSh-41 submachine gun, one of the weapons that was mass produced in the USSR during World War II that armed Soviet tank-riding infantry fighting on the Eastern Front.

make Soviet 7.62mm ammunition and posed an added logistic burden for Finnish front-line troops. However, the demand for more submachine guns remained high and, although Finnish army HQ explored the possibilities of modifying the gun to accept Finnish 9mm ammunition and Suomi magazines, production of this modified gun was never started.

The MP717(r)

The Germans themselves were very impressed by the PPSh-41, and would use the guns as often as they captured them. Since the 7.62mm and 9mm Parabellum cartridges shared similar dimensions, only a 9mm barrel and a magazine housing adaptor was needed to convert the PPSh-41 to fire from MP38/40 32-round magazines. The Wehrmacht (the German armed forces) officially adopted the converted PPSh-41 as the MP717(r).

The Leningrad gun

During the siege of Leningrad (1941–44), there were working munitions factories, but basic materials were in short supply. In these conditions A.I. Sudarev designed the PPS-42. It was soon improved and resulted in the Pistolet-Pulemet Sudareva 1943g or PPS-43 submachine gun, an all-metal weapon with a folding stock and a compact 35-round curved box magazine. This modern design remained in production

after the war. The folding stock reduced the weapon's length and made it ideal for tank crews, paratroopers, and reconnaissance units. While the weapon had a slightly slower firing rate of 700 rpm, it more than made up for this with its lighter weight, small size and greater ease of manufacture. About 500,000 were made. Captured PPS-43s were used by the Germans, with the designation MP719(r).

folding butt

35-round curved magazine

LEFT The PPS-43 had a folding stock that reduced its length to a compact 622mm/24.48in, making it an ideal weapon for Armoured Fighting Vehicle (AFV) crews and paratroops.

Submachine-gun design advances

Following World War II many countries produced submachine guns for their armed forces and police. Many looked at the design features of wartime weapons, and with the leisure of peacetime production refined and improved on these combat-tested submachine guns. Some, like the Swedish Kulsprutpistole Model 45, French MAT-49 and Czech Model 61 Skorpion took on what could be called cult status. In Italy the 9mm Beretta Model 12 submachine gun proved to be a reliable weapon that enjoyed a long operational life.

sight

hinged wire butt

magazine

ABOVE The Smith & Wesson Model 76, a licence-built version of the Swedish Carl Gustaf Model 45 9mm submachine gun. The US version was a popular weapon with the CIA in Vietnam in the 1960s and 1970s. It was also produced in Egypt by the Maadi Company as the Port Said.

"The Swedish K"
The Kulsprutpistole Model 45 submachine gun was developed by Swedish state-owned Carl Gustaf Arms company in 1945 and was manufactured under licence in Indonesia, and in Egypt under the name of "Port Said", and sold to Ireland.

An almost exact copy of the Model 45 was also produced in the United States by the Smith & Wesson company in the early 1970s as the S&W Model 76. The Model 45 was not adopted by the US navy, army or police forces, but known colloquially as "the Swedish K" it did gain fame or notoriety when the Central Intelligence Agency (CIA) and special forces used it in South-east Asia during the Vietnam War. A silenced version was used in covert operations. The CIA and other agencies were able to buy these weapons without going through US military sources and men in civilian clothes seen in Vietnam carrying a Swedish K were easily recognizable as "spooks" (secret agents or spies).

The weapon had an excellent 36-round two-column magazine that was very reliable and has therefore been copied or type developed in Czechoslovakia (as it was then), Scandinavia and Germany. Uncomplicated and reliable overall, loading and maintainance are generally problem-free. It is easy to field strip and clean, and has a quick-change barrel. Partly due to the "in-line" design of the stock the Swedish K is a good shooting weapon: it is balanced, easy to handle and control for

long bursts of fire. A compact gun, the rigid stock is folding and the cartridges push directly into the body, so the magazine does not need a loader.

The MAT 49
The French state arms factory MAT (Manufacture Nationale d'Armes de Tulle) developed the MAT 49 submachine gun in the late 1940s. Adopted by the French army in 1949, the MAT 49 was widely used by the French army and police for nearly 30 years, and it also saw action in Indo-China and Algeria. It is a solid, well-made submachine gun with a magazine that hinges forward to clip under the muzzle when the weapon is not in use. The retractable wire butt can be slipped forward to reduce the length to a compact 460mm/18in. During the Vietnam War the Viet Cong produced a local copy of the MAT 49, chambered for Soviet 7.62mm ammunition.

The Beretta Model PM 12 and PM 12 S
In 1959 the 9mm Beretta Model PM 12 submachine gun was developed by the venerable Italian small-arms manufacturer Pietro Beretta Spa. Designed by Domenico Salza, it went through a number of prototypes. The weapon was made from heavy sheet steel stampings spot-welded together to form the

tunnel foresight

retractable wire butt

grip safety

RIGHT The French MAT 49 has a 32- or 20-round box magazine that fits into a hinged housing. The telescopic wire butt is similar to that on the US M3 submachine gun. The MAT 49 fires at a cyclic rate of 600 rpm.

hinged housing and hand grip

receiver and magazine housing. The receiver, forward pistol grip, magazine housing, trigger housing and pistol grip are all one unit. It has two safety catches. The Model 12 was adopted by the Italian army in 1961 and sold under licence to Brazil. It was also sold in Gabon, Libya, Nigeria, Saudi Arabia and Venezuela. The company followed the Model 12 with the Model PM 12 S. The Model PM 12 S included a number of modifications and improvements, among which was an epoxy-resin finish resistant to corrosion and wear. The Model PM 12 S had a safety catch that when applied blocked the grip safety and trigger. Both weapons had a cyclic rate of fire of 550 rpm. The Model PM 12 and the Model PM 12 S were designed for mass production but are compact and well made.

Beretta PM 12 S
cal. 9 mm. Parabellum

ABOVE Promotional literature from Beretta for the PM 12 S. The compact 9mm submachine gun has a 32-round detachable box magazine with options for a 20- or 40-round magazine. This model has a folding metal stock, but it is also available with a wooden butt stock.

Skorpion's sting

The design philosophy behind the 7.65mm Model 61 Skorpion (Czech vz.61) was an almost futuristic concept in 1952. It was intended to replace the pistol, but could also be used as a close-combat assault weapon, as well as for personal defence. At 270mm/10.63in long with the butt folded, its small size makes it easy to conceal in a holster and can be used in confined spaces, such as cars or aircraft. Although it became popular with both police and security units in the 1960s, it was also used by terrorist groups.

The Skorpion has also been manufactured in Serbia as the Model 84, and police and security forces in Afghanistan, Angola, Egypt, Libya, Mozambique and Uganda have adopted it. The Skorpion was produced for export chambered for the 9mm Makarov cartridge and the Browning 9mm Short. The CZ91S is a semi-automatic version with very high "collector's" finish in black enamel.

folding butt

10- or 20-round curved magazine

ABOVE The Model 61 Skorpion was intended to be a dual-purpose weapon, combining the functions of pistol and submachine gun. Since it had a stock and can be held with two hands it was effective against targets that were beyond the range of a conventional pistol, and had greater firepower.

Invention and innovation

In 1950, Uzi Gal, an engineer with Israel Military Industries (IMI), developed a very effective 9mm submachine gun, which would become known as the "Uzi". The British L2A3 9mm Sterling submachine gun entered service with the British Army in 1957 and would see action in the numerous campaigns fought from Borneo to the Falklands. The compact American MAC 10 and MAC 11 submachine guns have proved popular with undercover formations around the world.

The Uzi

Since it entered service with the Israeli Army in 1951, the Uzi has been adopted by the police and armed forces of more than 90 countries, including Israel, Germany and Belgium. The more compact versions, the Mini and Micro Uzi, are used by many police, special operations and security units around the world, including the Israeli Sayeret and the US Secret Service.

The Uzi was based on the Czech M23 and 25 submachine guns, but with a completely different receiver (rectangular instead of round in cross-section) and other changes. It is a recoil-operated, select-fire submachine gun, firing from the open bolt (the bolt is the part of a firearm that blocks the rear of the

cocking handle

grip safety

hinged butt

ABOVE The Israeli 9mm Mini Uzi is a compact weapon that is ideal for close protection teams because it can be concealed inside a jacket. It can be fitted with a 20-, 25- or 32-round magazine.

ABOVE An officer "checks clear" to see that the breech is clear of rounds on the L2A3 Sterling submachine guns of a section of Gurkha soldiers during training in a Fighting In Built Up Area (FIBUA) complex in the 1980s.

Uziel Gal (1923–2002)

Uziel "Uzi" Gal was born Gotthard Glass on 15 December 1923 in the town of Weimar in Germany. When Hitler came to power, the ten-year-old Glass moved with his school to England. In 1936, he moved to Kibbutz Yagur in what was then the British Mandate of Palestine. In Palestine he joined Palmach, the underground infantry arm of the Hagana, working as an armourer.

In 1948 Gal began work on the Uzi submachine gun shortly after Israel became a nation. In 1951 the submachine gun was adopted by the Israel Defence Force (IDF) and though Gal did not want the weapon to be named after him, his request was ignored. In 1958, for his work on the Uzi, Gal became the first person to receive the Israel Security Award, presented to him by Prime Minister David Ben-Gurion.

cocking handle — sights

RIGHT The L2A3 9mm Sterling submachine gun entered service with the British Army in the 1950s. It was used around the world until the Gulf War.

magazine housing

folding butt

chamber while the powder burns). The bolt "sleeves" around the rear part of the barrel, reducing the overall length of the gun. The Uzi has a cyclic rate of fire of 600 rpm, while the Mini's rate is 1,500 to 1,900 and the Micro's is 1,200 to 1,400 rpm.

The Sterling submachine gun

The L2A3 9mm Sterling submachine gun entered service with the British Army in 1957 and soldiered on for almost 35 years. It would be used by officers, vehicle crews and radio operators in operations around the world until the Gulf War of 1990–91. It weighed 3,500g/123.46oz and was 690mm/27in long with its butt extended and 483mm/19in with it folded. The effective range was about 200m/654ft with a muzzle velocity of 390m/s (1,279ft/s) and cyclic rate of fire of 550 rpm. The standard magazine held 34 rounds, but was usually loaded with 28 to reduce pressure on the spring.

For covert operations stubby 10- and 15-round magazines, "stacked" for a quick change, were available. This arrangement proved efficient in use. Canada produced the Sterling submachine gun as the C1 and Spain as the C2.

American genius

When weapons designer Gordon B. Ingram returned to the United States at the end of World War II he began working on submachine gun designs. In the late 1960s he came up with the tiny MAC 10 and MAC 11 submachine guns built at his Military Armament Company (hence the name MAC) at Powder Springs, Georgia. Also known as the Ingram, the versions of the MAC 10 and MAC 11 without stocks are respectively 267mm and 222mm/10.5in and 8.75in long. If the wire stock is extended, the guns are 548mm/21.5in and 460mm/18.11in long. These compact guns have a phenomenal rate of fire: the MAC 10 fires at 1,145 rpm, the 9mm MAC 10 at 1,090 rpm and the 9mm MAC 11 at 1,200 rpm. These small submachine guns are popular with security forces on covert operations since their external thread muzzle can be fitted with a MAC suppressor.

BELOW The Ingram MAC 10 is in service with the US navy and several Central American countries but, despite being excellent weapons, the .45 and 9mm MAC 10 and 11 have never been widely adopted.

thread for suppressor

hinged butt

grip safety

magazine integral to grip

The Colt M4 carbine

One American weapon that falls between a rifle and a submachine gun is the Colt M4 Carbine, the most compact of the Colt range of 5.56mm weapons. It features a four-position telescopic butt, a barrel length of 292mm/11.5in, and is designed for use wherever lightness, speed of action, mobility and close-quarters combat are required. It uses the same 30-round magazine as the M16 rifle. One veteran of the Vietnam War carried an M4 Carbine that had developed a reputation as a "jinxed" weapon since its previous users had been killed. Far from being jinxed, he said, it was a very handy weapon. It did, however, have a different signature when it was fired and consequently had attracted hostile fire when its former users had been caught in a firefight. As he explained, what had killed them was not the weapon, but their field craft and low-level tactics.

Assault rifles to submachine guns

One of the most influential submachine guns of the post-war period was the Heckler & Koch MP5, but it was rivalled by the Soviet AKSU-74, a modified AK-74 assault rifle. In Chile, Fabricas y Maestranzas del Ejercito (FAMAE) produced a modular range of submachine guns under the designation SAF, and during the 1970s Russian Evgenij Dragunov designed the Kedr, a submachine gun that departed from the AK format.

The HK54

The West German Heckler & Koch MP5 submachine gun was first produced in the mid-1960s as the HK54. It was designated Machine Pistol 5, (MP5), when it was adopted by the West German government for use by its police and border guards. It was a lightweight, air-cooled, magazine-fed, select-fire weapon with delayed blowback operation. It could be shouldered or hand-fired in semi-automatic, with two- or three-round bursts, and sustained fire modes. It is a very accurate weapon that has a cyclic rate of fire of 800 rpm.

The modular design of the weapon consisted of six assembly groups and provided an unmatched degree of flexibility, as these groups can be exchanged with optional parts to create various styles of weapons for

tactical light

retractable butt

magazine

ABOVE The Heckler & Koch MP5 mechanism is used in six different submachine guns including three silenced versions and the compact easy-to-conceal MP5K that is popular with counter-terrorist teams.

numerous operational requirements. The metal surfaces of the MP5 were phosphated and coated with a black lacquer finish that is highly resistant to salt-water corrosion and surface wear. The MP5 is made or has been made under licence in the United States, UK, Turkey, Pakistan, Mexico, Iran and Greece.

The AKS-74U

Modified from the Russian 5.45mm AK-74 assault rifle, the AKS-74U was specifically designed for vehicle crews who needed a weapon with a much shorter barrel. The overall length of the submachine gun is only 492mm/19.37in with the metal stock folded, or 728mm/28.66in with extended stock. The AKS-74U has other advantages over the AK-74 assault rifle, as its loaded weight of 3,100g/109.34oz makes it significantly lighter and it also has a somewhat higher cyclic rate of fire at 650 to 735 rpm. The rear sight is a flip-type U-notch, while the front sight is a cylindrical post. At the end of the barrel, there is a device which operates as an expansion chamber to bleed off gases that could otherwise cause a violent recoil. It is also produced in Poland as the ONYX and in Serbia as the Zastava M85.

ABOVE The AKSU-74 has become notorious as a weapon favoured by terrorist and dissident groups in Afghanistan. Its compact length of 492mm/19.37in makes it an ideal weapon for such purposes as it fires powerful 5.45mm rounds from a 30-round magazine.

rear sight

cocking handle

sling swivel

hinged butt

magazine

ABOVE The Spanish Star 9mm Model Z-84 submachine gun has a cyclic rate of 600 rpm. It is constructed using stampings and investment castings that helps to speed production and keep costs down.

Compact Chilean

Fabricas y Maestranzas del Ejercito or FAMAE, the Chilean arms manufacturer, took the Swiss SIF 540 assault rifle that they were manufacturing under licence and developed it into the SAF and Mini-SAF submachine gun. The 9mm submachine gun comes in several forms: the standard with a fixed butt stock, standard with folding stock, silenced and Mini-SAF. They employ 20- or 30-round box magazines, and the 30-round is made from translucent plastic so a quick visual check can establish how much ammunition is available. The SAF submachine gun is in service with Chilean armed forces and police.

Russian police weapons

In the early 1970s, Evgenij Dragunov, the designer of famous SVD sniper rifle, designed the Kedr submachine gun (then known as the PP-71). The project was shelved, but was revived in the early 1990s. The PP-71 was slightly modified and went into

production in limited numbers, and was issued to Russian police units. New, more powerful ammunition was developed to replace the original 9mm Makarov ammunition. This new round, the PMM, while retaining the same dimensions, had a slightly lighter bullet and heavier charge, which increased its performance. In 1994 the Kedr was slightly strengthened to take this new ammunition, and the weapon was marketed as the submachine gun.

The Spanish Z-84

Despite being compact and lightweight, the Spanish Z-84 is a powerful submachine gun which was developed in the mid-1980s by Star Bonifacio Echeverria SA. It fires soft point and semi-jacketed bullets, as well as full metal jacket military ammunition, from 25- or 30-round magazines. Once it has been cocked the submachine gun has no external moving parts, which is one of the reasons that it has been adopted by special forces in the Spanish army and police units.

Guns of the Special Air Service (SAS)

When six armed revolutionaries besieged the Iranian Embassy in London on 30 April 1979, Prime Minister Margaret Thatcher authorized the use of elite SAS counter-terrorist troops to neutralize them.

On 5 May, two SAS troopers clad in hooded black overalls were spotted outside the embassy as they positioned a frame charge against an upper-floor window. The large explosion that followed ten seconds later was a signal for the assault by teams at the front and rear, who moved nimbly across the balconies and through the smashed window.

The SAS weapon was the Heckler & Koch MP5, which they had encountered during joint training with the West German GSG9 counter-terrorist force, and which the SAS nicknamed the "Hockler".

ABOVE SAS troopers armed with Heckler & Koch MP5 submachine guns storm through the blasted windows at the front of the Iranian Embassy at Princes Gate, London, on 5 May, 1980.

Silenced guns

In World War II Britain pioneered some silencer concepts, notably with the Sten Mark VI (S) which was the world's first silenced submachine gun when it was introduced in 1943. Britain also produced the hybrid De Lisle Silent Carbine for use by covert forces. After the war the British developed the L34A1 silenced Mark 5 Sterling submachine gun. The Chinese made two rather bulky silenced pistols in the 1960s: the Type 64 and Type 67.

Wartime innovations

The Sten Mark VI (S) submachine gun was intended for single shots only, though it could be fired on automatic in an emergency. With the silencer the weapon was 908mm/35.75in long, compared to the 762mm/30in of a standard weapon. It was also popular with Resistance groups, and German counter-intelligence officers were greatly impressed by captured weapons. Silenced weapons have been improved since 1945 so that the silencers have a longer life even if the gun is fired in bursts.

ABOVE The Sten Mark VI (S) submachine gun was almost completely silent. It had a lower muzzle velocity compared with conventional Stens and lower cyclic rate of 450 rpm to reduce wear.

De Lisle Carbine

The British De Lisle Silent Carbine is hard to pigeonhole – is it a bolt-action pistol or a pistol-calibre rifle? What it definitely is, is a silent weapon. To work effectively with a silencer a weapon has to generate a muzzle velocity below the speed of sound, around 330m/s (1,082.67ft/s) depending on altitude and air temperature. The De Lisle Carbine, developed for British Commandos and special forces in World War II, took a .303 Lee Enfield bolt-action rifle but chambered it for the standard .45 ACP from the Colt 1911 pistol. A massive silencer gave it a 210mm/ 8.25in barrel. The first examples used the standard

Lee-Enfield wooden stock but a paratroop version with a folding stock was also produced. The signature or noise when the De Lisle Carbine was fired was a quiet "plop", a sound that would not arouse suspicion. However, although the carbine was reportedly accurate to over 300m/984.25ft, the velocity of around 260m/s (853ft/s), combined with a relatively light round-nosed bullet, produced a curved trajectory at long range, making accurate range estimation crucial.

The L34A1 Mark V Silent Sterling

The British L34A1 Mark V Sterling was a silent version of the 9mm L2A3 Sterling. It weighed 4,300g/151.68oz, and was 864mm/34in long with the butt extended. It had an effective range of 152m/500ft, with a muzzle velocity of 292–310m/s (958–1,017ft/s). The basic

LEFT The Silent Sterling L34A1 uses an ingenious silencing system that allows gases to escape through radial holes into an expansion chamber and then through a series of baffles. It fires standard 9mm cartridges, unlike other silenced weapons that require subsonic ammunition.

bolt action

adjustable sights

magazine release catch

sling swivel

mechanism was the same as the Sterling but the Silent Sterling has a fixed silencer allied to a special barrel. In the barrel, 72 radial holes are drilled through allowing for the hot propellent gases that follow the bullet out of the muzzle to enter an "expansion chamber". Here these gases expand to dissipate some of their energy. The gases pass through a series of metal baffles, with a central hole to allow the passage of the bullet. The baffles deflect and slow the flow of gas emerging from the expansion chamber. By the time the gases emerge from the silencer, they are cooler, slower and silenced.

Compared to the standard Sterling the L34A1 Mark V appears clumsy and unbalanced. In fact, while it is not as compact, it handles well since the silencer is not intrinsically very heavy. However, it is recommended the silenced weapon should only be fired on single shots and automatic used only in an emergency. The standard Sterling submachine gun is not a noisy weapon when it is fired, but with the

ABOVE The De Lisle Carbine made by the Royal Small Arms Factory at Enfield, with a ten-round magazine. It was fitted with a .45 barrel forming part of a large, and very efficient silencer.

silencer fitted it is almost inaudible at 30m/100ft. Before the Heckler & Koch MP5 SD became more widely available on the world market, the L34A1 Mark 5 was very popular. Ironically, while it was used by British Special Forces, it was also bought by Argentina. The L34A1 Mark V was issued to the Special Forces who spearheaded the invasion of the Falkland Islands in 1982.

Chinese silenced pistols

China had initially looked to the Soviet Union for weapons and copied these designs. However, in the Type 64 and its successor the Type 67, Chinese small arms designers came up with ingenious silenced weapons. The nine shot Type 67 and 64 pistols are built around an integral silencer. The Type 64 silencer is a light steel cylinder that contains steel mesh and several baffles that slow down and cool the gases from the muzzle blast. It could not fire standard ammunition. Instead it fires a relatively low-powered 7.65mm ammunition developed for the weapon that has a muzzle velocity of 241m/s (790ft/s). Though this helps to keep the sound down, the penalty is a short effective range of between 9–18 m /30–60ft. The pistol can also be fired manually using a lock system that prevents the semi-automatic action from operating.

The Type 67 silenced pistol is a further development of the Type 64 and has replaced the Type 64 in Chinese service. It provides the users with the same combat characteristics, but weighs 1,050kg/37.04oz in contrast to the 1,810g/63.85oz of the Type 64, and the shape of the silencer has been changed to make carrying easier.

Silent Sterling

There are few times when British Special Forces are on public view – and fewer still of the L34A1 Mark V Sterling. Ironically, the greatest exposure that the weapon received was on the morning of 2 April 1982 on news reports. It was slung from the shoulder of Sergeant I.M. Batista, the Argentine commando who escorted the Royal Marine naval party who had surrendered after their overnight defence of Government House in Port Stanley on the Falkland Islands. Batista's imperious manner and the glum Royal Marines with their hands raised in surrender did much to galvanize public opinion in Britain to back the despatch of a Task Force to the Falklands.

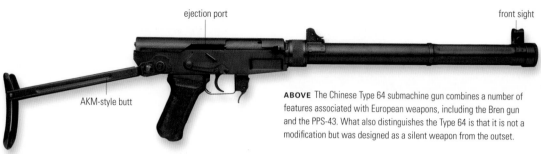

ejection port   front sight

AKM-style butt

ABOVE The Chinese Type 64 submachine gun combines a number of features associated with European weapons, including the Bren gun and the PPS-43. What also distinguishes the Type 64 is that it is not a modification but was designed as a silent weapon from the outset.

Chinese submachine guns

Though the Chinese Type 64 was one of the first indigenously designed weapons it incorporated a number of features taken from European weapons. The bolt action comes from the Soviet PPS-43, the folding butt from the AKM and the trigger mechanism from the British Bren LMG. The silencer is integral to the SMG. It is said to be reasonably effective and to reduce any muzzle flash, and the use of the subsonic 7.62mm Type 64 round further reduces the firing signature. The gun can fire conventional 7.62mm TT ammunition but the higher velocity wears out the silencer. For a submachine gun, the Type 64 has a surprisingly high rate of fire at 1,315 rpm – however, it is a selective fire SMG and would probably be used for single shots.

The Type 85 produced by China North Industries Corporation (NORINCO) is a simplified and lighter version of the Type 64 that has been produced for the export market. The silencer is similar to that on the Type 64.

It is most effective firing Type 64 ammunition with its heavy, subsonic bullet, but it can also fire standard 7.62mm ammunition, although there is an increased noise level. The Type 85 uses the same 30-round box magazines as the Type 64 submachine gun.

Silenced German submachine guns

The German Heckler & Koch MP5 SD submachine gun series (SD1–SD6), introduced in 1974, was a superb weapon widely favoured by Western special forces. It had an integrated suppressor and a specially made barrel that has 30 2.5mm ports that reduce the muzzle velocity of its ammunition to subsonic and the the bullet crack and muzzle blast are much reduced. The result is that the MP5 SD series is almost inaudible at distances of more than 15m/49.21ft. Because of this remarkable sound reduction and the almost invisible muzzle flash, the gun has been selected by many police and military units around the world.

Silent Soviet killers

The Soviet two-shot MSP pistol does not use a silencer, but instead has an ingenious type of ammunition, making it well suited to close-range assassination operations. The pistol was developed around 1972, intended as an easily concealed, short-range assassination weapon for the KGB, GRU, Spetznatz and other covert organizations. Because of its specially designed SP-3 silent ammunition, which encapsulates all powder gases within the cartridge case, the pistol does not require a silencer.

The MSP pistol is a two-barrel, non-automatic pistol. Two barrels are in an "over-under" configuration, and are tipped up at the rear for loading and unloading. The two cartridges come clipped together which makes reloading quicker. The trigger is single-action, with the manual safety catch just behind the trigger on the left side of the grip. The enclosed hammers are cocked manually using the cocking lever at the bottom of

ABOVE A silenced MP5, part of the series SD1–6. The silencer requires no maintenance except rinsing in an oil-free cleaning agent.

The sound of silence

No automatic weapon fired in bursts is completely silent, and Heckler & Koch have the correct description when they describe the MP5 SD as *Schalldämpfer*, literally "sound dampened".

When a gun is fired there are actually two noises – the explosion of the cartridge and the high-velocity crack of the bullet if it breaks the sound barrier. In a firefight, timing the interval between "crack and thump" gives both an indication of range and also whether the rounds are being aimed at the listener. If there is a distinct delay between the crack and thump then the weapon is being fired at long range and consequently the shooting may not be very accurate. If, however, the sounds almost run together the marksman is nearby and this is the time to be worried. If all that is heard is the thump of the exploding cartridge the rounds are being fired in a different direction.

A suppressor, widely called a silencer, will reduce the speed of the bullet so that it is travelling below the speed of sound, thus removing the "crack" noise. The thump of the exploding cartridge can also be reduced so the sound is something like that of a pneumatic stapling gun or the rattle of machinery.

RIGHT A "night vision" view of a special forces operative clearing a room using a silent submachine gun, the MP5 SD, to dampen the sound in an enclosed area.

the trigger guard. Surprisingly, given that engagement ranges would probably be measured in a few feet, the pistol has small open sights.

Designed in the mid-to-late 1970s, the Soviet S-4 and S-4M pistols further developed the concept of the MSP, and were used in Afghanistan, as well as in many counter-terror operations closer to home. However, possibly because it was seen as a military weapon, the bigger S-4 pistol sacrificed ease of concealment for greater range and hitting power from the powerful silent PZ cartridges.

Adopted around 1983, the Soviet 7.62mm PSS Pistolet Spetsialnyj Samozaryadnyj or Special Self-loading Pistol is a sophisticated design that is a significant improvement on the MSP and S-4M silent pistols in performance and particularly firepower.

It has a six-round box magazine and is much more compact and silent in action than conventional silenced pistols like the Soviet PB or Chinese Type 67. Currently, most elite Russian anti-terrorist teams use the PSS.

RIGHT The Russian PSS is a compact silent weapon that has a six-round magazine. The special SP-4 ammunition has an internal piston that encapsulates the propellent gases in the cartridge case. The breech block can also be locked closed to prevent the empty case being ejected to rattle to the ground, so ensuring complete silence.

Specialist and personal defence

Some Personal Defence Weapons (PDWs) are almost novelty guns, while others, such as the classic Belgian FN P90 5.7mm, have a useful role as back-up weapons and are compact enough to be carried and concealed easily. The Personal Defence Weapon falls somewhere between a submachine gun and a handgun, and the Chinese and Soviet Union have both developed gun/knife combinations, an idea that actually dates back to the Frank Wesson dagger-pistol from the 1860s in the United States.

Guns and knives

Introduced around 1994, the Chinese Firing Combat Knife falls outside the PDW category but merits description. Its blade is 40mm/1.58in long and it has four barrels chambered for .22 LR rounds built into the cylindrical handle. The barrels are positioned so that there are two on either side of the blade, with the muzzle at the base of the blade. By pressing a spring-loaded cap at the rear of the grip, the hilt cap is removed and rounds can be inserted into the chambers. The safety catch is a serrated ring in front of the handle. If it is rotated clockwise this locks the trigger, which doubles as a handguard. With the catch on, the combat knife can be used in knife mode. Twist the safety catch to the "off" position and the Combat Knife is ready to fire. The Firing Combat Knife has an effective range of 18m/60ft, although it is probably more useful at shorter ranges.

Although the Russian 7.62mm Scouting Knife fires only one round, this is a 7.62mm silent cartridge. Unlike the Chinese weapon, the muzzle of the Russian knife is in the hilt – so to fire it the knife is reversed and fired by pressing the trigger bar in the handle. A sliding safety catch prevents accidental firing. The effective range is 23m/75ft. The blade of the Scouting Knife is electrically insulated and is 152mm/6in long and 25mm/1in wide – substantial enough to cut heavy cables. A screwdriver is also incorporated.

Small but deadly

The Heckler & Koch MP5K introduced in 1976 was a dumpy, short-barrelled version of the MP5 that had no shoulder stock and used a small 15-round magazine. It was adopted by German special forces, and police and military forces throughout the world.

In 1991 H&K in the United States produced a version with a special hinged shoulder stock that was renamed "MP5K-PDW," which could fire from 15- or 30-round magazines. Only 368mm/14.5in long with the butt folded, it became a much more stable weapon to fire from the shoulder or the hip once it was extended.

The Finnish 9mm Jati-Matic/GG-95 PDW has an unusual barrel location that allows the bolt to recoil up an inclined plane at an angle to the barrel and so reduce muzzle climb when fired on automatic. It has a select-fire trigger and a (rudimentary) folding foregrip. It fires from a 20- or 40-round box magazine and has a cyclic rate of 600 to 650 rpm. Although it is very compact, the fact that it fires 9mm ammunition puts it up against more proven systems like the Hecker & Koch MP5K-PDW. Despite this the Jati-Matic is manufactured under licence in China.

ABOVE Due to its compact size the MP5K has high mobility. However, it is not as popular as the PDW version of the MP5K which has a shoulder stock making control of the weapon much easier.

ABOVE Packed ready for transport, the SPP-1M Russian underwater pistol. It can be fired underwater, and on land when the diver has reached the shore, making it a truly amphibious weapon.

Combat beneath the waves

The Russian Tsniitochmash organization produced the *Spetsialnyj Podvodnyj Pistolet* – the Special Underwater Pistol or SPP-1 – in the late 1960s for combat divers working in the Soviet navy. The SPP-1 is a four-barrelled, non-automatic weapon that fires 115mm/4.5in-long drag-stabilized darts that are lethal at 45m/147.5ft in the air and 15.5m/51ft under water. It is also lethal at greater depths than this, although at closer ranges. The sealed ammunition is loaded from the rear and the self-cocking mechanism fires one round with each pull of the trigger. The SPP-1 has been modified and upgraded as the SPP-1M and is still used by the Russian navy special forces.

The definitive PDW

The Belgian P90 5.7mm PDW designed by Fabrique Nationale de Herstal (FN Herstal) entered service in 1994. It is a truly innovative weapon built around a new FN round that is a cross between a rifle and pistol bullet. PDWs are now required in service because existing pistol calibre cartridges are becoming increasingly ineffective at disabling troops equipped with body armour.

It is a lightweight and completely ambidextrous weapon with a compact, translucent 50-round magazine that runs horizontally across the top of the P90, and empty cases are ejected downwards, well away from the operator. Only the bolt and barrel are made of steel, the rest being from high-impact plastic material. It fires from a closed bolt, giving it outstanding accuracy and making it easy to keep on target. It has a cyclic rate of fire of 900 rpm and a combat range of 137m/450ft.

BELOW The FN P90 fires a new type of 5.7mm ammunition that uses dense plastic in place of conventional lead and can penetrate 48 layers of body armour, making it particularly lethal.

Magnum force

Although the compact convenience and large magazines of automatic pistols seemed to spell the end of the revolver, the weapon still had its advocates after World War II and was much favoured by police. These guns include the .44 American Smith & Wesson Magnums, the .357 Colt Python and the .357 Ruger Blackhawk Magnum. The Russian 12.3mm Udar is a different type of revolver, but it still delivers a formidable punch.

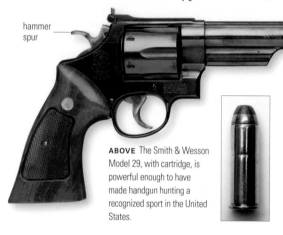

hammer spur

ABOVE The Smith & Wesson Model 29, with cartridge, is powerful enough to have made handgun hunting a recognized sport in the United States.

Classic revolvers

One American handgun, the S&W Model 29, would become something of a film star in the 1970s. From 1955 to 1957 the big Smith & Wesson revolver was simply called "the .44 Magnum", but when in 1957 S&W standardized the model numbering of their products it became the S&W Model 29. Smith & Wesson .44 Magnum revolvers have heavyweight steel frames produced in blue or stainless-steel finishes. The blue and nickel models are called the Model 29, while the stainless-steel version is the Model 629. Versions such as the "629 Classic", "629 DX", "629 Classic Hunter" have specialized features such as interchangeable front sights, full lug barrels, and special grips. *Jane's Infantry Weapons* sums up the reputation of the .44 Magnum: "When the utmost power is required from a handgun, the Model 29 is the obvious answer, delivering upwards of 1600 J of muzzle energy with the longer barrels."

The revolver revival

When it was introduced in 1955, the hand-made Colt Python .357 revolver immediately became one of the most popular personal weapons for US Law Officers.

It was available either blued or nickel plated, however, the nickel plated finish was replaced by satin or polished stainless steel finish. Uniformed officers carried the service revolver with its 101mm/4in or 152mm/6in barrels, while their colleagues in plain clothes tucked the self-defence versions with their stubby 64mm/2.5 in or 101mm/4in barrels underneath jackets. Hunters and target shooters opted for the big 101mm/4in and 203mm/8in versions.

The weapons were well made, comfortable to fire and since they had originally been intended to be a big calibre target revolver – very accurate. The foresights were conventional with a red polymer dot insert, the rear sights adjustable for both windage and elevation and had a white edging. Production stopped in 1998 when the Python was replaced by the Kingcobra, however, individual revolvers were made until 2003.

ABOVE The discharge of a Magnum revolver at a shooting range. The target is aimed at using an optical scope. Large-calibre weapons are best suited to target shooting and hunting.

In 1955 the American gunmaker William Ruger, who had noted popular interest in single-action revolvers seen in Western films, produced a single-action gun that quickly became the flagship for all his single-action handguns – the .357 Blackhawk Magnum. This was followed by the .44 Magnum Blackhawk in 1956. Three years later the largest of the single action Rugers appeared, the Super Blackhawk, which had an all steel frame and could fire either .44 Magnum or .45 Colt ammunition. However, Ruger revolver designs were also innovative. The compact SP 101 is a stainless steel five shot weapon in various calibres and capable of handling powerful modern ammunition. In 1993 a double action pistol with a spurless hammer was introduced that was ideal for covert use as there was no risk of the spur snagging on clothing. It was available in 0.38 Special and 0.357 Magnum chambering.

ABOVE The Colt Python was originally made in a nickel-plated finish. This was changed when stainless-steel became available, supplementing the existing blued carbon-steel.

The Udar revolver

If a modern version of the film *Dirty Harry* were produced, the hero might choose the Russian 12.3mm Udar or "Blow". This compact weapon is a double- and single-action revolver designed to fire a range of special cartridges built around the 32-gauge hunting round, and has been developed for police and security forces. Clips are preloaded with different ammunition types that can be quickly replaced according to the tactical situation.

The armour-piercing round has a sub-calibre steel bullet that can penetrate a 5mm/0.2in steel plate at 45m/147.64ft, as well as conventional body armour. A plastic round is available for non-lethal operations, and a pyroliquid (coloured ink) bullet for training exercises and for marking offenders. The pistol fires conventional ball ammunition and a high-noise blank round. Empty cases are ejected simultaneously. Three filled reserve clips are carried in the box, together with the holster.

"Do you feel lucky?"

Don Siegel's 1971 film *Dirty Harry* was considered sensational because of its overt violence. The duelling combatants, Clint Eastwood's unconventional cop and the pathological sadistic criminal, share traits of brutal violence and insanity.

When the film was being made, the .44 Model 29 Magnum was not in production, so strings had to be pulled to get a few made especially for the film. Eastwood spent time on the range prior to undertaking the role, so that he could properly replicate the .44's recoil when firing blank cartridges in the film. The blanks also had to be specially made, as the traditional Hollywood blanks would not fit the .44 chamber. In the film, Harry uses a light special load that gives better control because of its reduced recoil.

RIGHT Clint Eastwood in *Dirty Harry*. In the film he describes the S&W Model 29 as "The most powerful handgun in the world". This, together with "Do you feel lucky, punk?", are among the memorable lines uttered by the ruthless detective in the film.

Post-war pistols

Pistol design expanded vastly after World War II in Europe, when the major users were police and the armed forces, with only a small number of private owners. In the United States, however, there was a large domestic market, too. Many designers took the best features of existing pistols, notably the German P38 and FN Browning 9mm High-Power and incorporated them into their pistols.

ABOVE The Czech 9mm CZ75 is a handy weapon that incorporates features like the safety from the P38 and the double stack magazine from the Browning HP.

French service
The French 9mm MAS Model 1950 pistol was developed in the late 1940s by Manufacture d'Armes de Saint Etienne (MAS) in France, and manufactured by MAS and also by Manufacture d'Armes de Chatellerault (MAC) until 1970. The pistol was a development of the pre-war SACM Model 1935A pistol, designed by Swiss engineer Charles Petter. This recoil-operated, locked breech, semi-automatic pistol, uses a single-action trigger with slide-mounted safety. This locks the firing pin, which, by pressing the trigger with the safety engaged, enables the hammer to be lowered. The MAS Model 1950 has fixed sights and a single-stack nine-round magazine.

Czech copies
Although the 9mm CZ75 pistol developed by the brothers Josef and Frantisek Koucky, seemed an ideal military pistol, it was only adopted by police forces. It would, however, prove to be a very influential design. It first appeared in 1975, with production beginning a year later. Obviously intended for the export market, it looked good, was comfortable to handle and shoot, and was quite accurate and reliable.

The CZ75 design had drawn on the best features of previous pistols such as the P38 and the Browning High Power, and in turn was widely copied and cloned. Israel's IMI produced the Jericho-941; in Italy Tanfoglio made the TZ75, TZ90 and T95; in Turkey Sarsilmaz manufactured the M2000; in Switzerland ITM the AT-88; while in China Norinco designed the NZ-75.

Spanish designs
Two years after Llama-Gabilondo y Cia of Vitoria produced their 9mm M-82 pistol it was adopted by the Spanish Army. It has a 15-round magazine and is a recoil operated, locked breech pistol with a short recoiling barrel like the Walther P-38 or Beretta 92. To load with the safety "on", a magazine is inserted, the slide operated, a round is fed in, but the hammer drops in a safe condition. To fire, push the safety to "fire" and pull the trigger – this cocks and releases the hammer. For greater accuracy, the hammer can also be manually cocked for the first shot. With a steel frame and slide, the M82 is relatively heavy at 1,110g/39.15oz, but it is also strong; an alloy frame version weighs in at 875g/30.86oz.

ABOVE The Spanish Star Model 30M has a 15-round magazine and an ambidextrous safety catch. The trigger guard is shaped for a double-handed grip.

The 9mm Spanish Star Model 30M and 30 PK was developed and manufactured by Star Bonifacio Echeverria SA. It appeared in 1990 and is now in service with the Spanish army and police. An updated version of the earlier Star 28M, they are recoil-operated, locked-breech pistols, that use Browning-designed, linkless locking. Safety features include a magazine safety that prevents firing without a magazine, as well as a chamber loaded indicator.

Safety and weapons handling

ABOVE The pistol is safe when the breech is empty of ammunition and the safety catch applied.

A firearm is a lethal tool that requires very little skill to use. To prevent accidents there are two simple safety rules. The first is to ensure that a weapon is unloaded. Remove the magazine while pointing the weapon in a safe direction. Then operate the action, which will eject a round that might be in the breech. Then, and only then, is the trigger pulled.

The weapon can then be "made safe" with the full magazine in place, but with the safety catch applied and no rounds in the breech. If it is unloaded, the magazine is removed and any rounds that may have been ejected are stowed back in the magazine, which is then put in a pocket or pouch.

If a weapon is handed from one user to another, he should, in British military parlance, "show clear", that it has been unloaded. A revolver is broken to show the empty cylinders, while the magazine is removed from an automatic weapon and the action pulled to the rear to show that the breech is empty. The second safety rule is never to point a weapon at anyone in fun.

In 1993 the Heckler & Koch Universal Self-loading Pistol (USP) – was marketed initially in 9mm and .40 S&W. With a modified Browning linkless locked-breech action with patented recoil reduction system, it is a very accurate pistol to fire since it is easy to recover the sight picture between shots. The USP has a moulded polymer frame with special grooves for the quick mounting of laser-aiming modules or tactical lights.

Italian style on trial

Beretta 92 pistols first began production in 1976 and were based on the early Beretta model 1951. Developed by the Italian company Armi Pietro Beretta Spa, the Model 92 was adopted by the Italian army as well as being manufactured under licence in Egypt. It is a blowback operated, locked-breech semi-automatic pistol which has a vertically tilting locking block system and an external trigger bar on the right side of the frame similar to the Walther P38. It also has a double-action trigger with a frame-mounted safety. Aluminium alloy is used for the frame and steel for the slide. Although the original Beretta 92 is no longer being made, it has served as the basis for numerous modifications and improvements.

In 1985 the Beretta model 92F was submitted as the XM9 for US Army Pistol Trials. Improvements and modifications included a chromed barrel, a trigger guard designed for a double-handed grip, new grip panels and a trigger bar disconnect safety. There was considerable controversy in the United States when it was finally adopted as the 9mm M9 pistol. It is manufactured in Italy, and in the United States by Beretta United States.

ABOVE The Italian Beretta 92 was adopted by the US Army as the M9. It is available in several configurations, all of which have a Browning HP-style 15-round magazine.

New guns, new materials

The attraction of new materials such as alloys and glass-reinforced plastic (GRP) in weapons construction is that they are lighter, sometimes stronger and do not rust or corrode. With a lower metallic content, they can be harder to detect by conventional security systems (found at airports for example), although some weapons developed using these new technologies have attracted controversy in the trade over materials and design.

The Glock

Glock Gmbh of Austria had an established reputation for its combat knives and entrenching tools when its 9mm Glock 17 won at the Austrian army pistol trials. The gun was subsequently adopted in the early 1980s under the designation P-80 by Austrian police and military forces. Since then, the Glock 17 and its subsequent derivatives have become very popular firearms, being exported to more than 50 countries. Among the features that have attracted users are the small number of working parts (only 33 including the magazine), and the use of highly resistant polymer materials. The Glock was probably the first handgun to be made successfully using polymers.

Currently, Glocks are chambered for all major pistol calibres, and are available in full-size service models, semi-compact models, compact models for concealed/backup use, and in longslide competition models. The Glock Selective-Fire Submachine Pistol has fitted a fire selector and 19- or 33-round magazine to the basic pistol and produced a weapon with an awesome rate of fire of around 1,300 rpm. It is possible

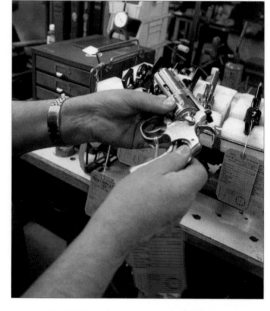

ABOVE A Colt .357 Python revolver receives the hands-on treatment on the production line at the factory at Hartford, Connecticut, United States. The company introduced the pistol in the early 1960s. While new guns and materials are introduced, older versions are still popular.

to obtain training versions, which fire non-lethal practice ammunition. To avoid confusion, the frame colour of "live" guns is red while guns that fire non-lethal ammunition are coloured blue.

Special operations

The Heckler & Koch Mark 23 SOCOM pistol was developed to a requirement from the US Special Operations Command for a pistol of a calibre of no less than .45 with a 12-round magazine. The gun must have a service life of 30,000 rounds before major maintenance, and be capable of being fired when wet,

ABOVE The revolutionary Glock 17 pistol has been adopted by the Austrian army and the forces of 45 other countries.

ABOVE The Heckler & Koch Mark 23 originated in 1991 when Heckler & Koch began development of a pistol for the US Special Operations Command (SOCOM). It finally passed all their rigorous tests and was issued by SOCOM in 1996.

dirty, icy or not maintained. A silencer and laser-aiming device were added to the specifications, which had to be instantly removable without tools and which would not adversely affect the accuracy of the gun upon refitting. Both Colt and Heckler & Koch entered the trials.

The H&K pistol had several parts (the slide and 12-round box magazine) made from polymers, and a mechanical recoil-suppression system that reduced recoil forces by 30 per cent. As well as the required flash and noise suppressor, it also had a laser-aiming module that could be attached to give an infrared or visual aiming spot. The H&K gun finally passed all the tests and became the .45 MK23 SOCOM offensive handgun and was issued in 1996.

Police and civilian guns
The 9mm Vektor Compact pistol (CP1) manufactured by the DENEL Corporation of South Africa was another weapon built using a polymer frame and a stainless-steel barrel. Introduced during the late 1990s, it was intended as a concealed-carry weapon for police and civilian use. Magazines are double-stack and the Compact's 12-round magazine fits flush with the bottom of the grip, while the 13-round has extended finger rests at the bottom.

Weighing 720g/25.4oz with the 13-round magazine, the CP1 seemed a useful little handgun. However, controversy struck when the US gun enthusiast press published a recall notice in October 2000 warning that the loaded pistol could fire if bumped or dropped, and advised users not to load the gun under any circumstances.

Hollow point or full metal jacket (FMJ)

The name "dum-dum" was used to describe a British military bullet that was developed at the Dum-Dum Arsenal, on the North-West Frontier of India in the late 1890s. It consisted of a jacketed .303 rifle bullet with the jacket nose open to expose its lead core. The aim was to improve the bullet's effectiveness by increasing its expansion upon impact; what this actually meant was terrible injuries to the victim. The phrase "dum-dum" was later expanded to include any soft-nosed or hollow-pointed bullet. The Hague Convention of 1899 outlawed the use of these bullets.

However, although soldiers are required to fire bullets that have a full metal jacket (FMJ), the police are not. While hollow-point rounds can make big holes when they hit human tissue, the rounds simply flatten if they hit a building, and will not ricochet off with the consequent chance of injuring innocent people in the vicinity.

ABOVE The hollow-point round causes extreme expansion or disintegration on impact.

Controversy
In 1994 Smith & Wesson introduced the Sigma series of pistols with the Sigma 40F in .40 S&W, and this was followed by a 9mm version. S&W had taken ten years to develop the pistol, which was the first to be made using polymer for both the gun frame and slide. However, the Sigma pistol was so similar to a Glock, that some referred to the Sigma series as "Swocks", a play on words between S&W and Glock. Glock sued Smith & Wesson, and although Smith & Wesson were obliged to pay an undisclosed sum to Glock for breach of patents, they received the rights to continue production of the Sigma line, with some amendments.

Modern handgun mechanisms

By the early years of the 20th century, the semi-automatic pistol and revolver had established themselves as viable weapons although the basic principles have not changed greatly since the 1930s. There are dozens of different operating methods for firearms, but the same basic principle applies to all: the firearm must be loaded, the hammer cocked for firing, and the firing pin must strike the primer of the cartridge for it to fire.

Single-action mechanism

Single-action (SA) pistols have a single-action trigger mechanism. The hammer is cocked and the trigger fired to discharge a bullet. A semi-automatic pistol such as the Colt M1911 is an example of a pistol that must be cocked for the first shot, but after the first shot has been fired, it is cocked automatically. These types of pistols typically have a very light trigger pull, and are very accurate. The pistol mechanism is always partially cocked while being carried and during firing, but cannot discharge a cartridge before it is fired.

Double-action mechanism

Traditional double-action (TDA) pistols have a mechanism that can be used in two ways. They can either be pre-cocked, like the single-action gun, or they can be fired with the gun uncocked. In this case, the gun has an additional mechanism added to the trigger that will cock the gun as the trigger is pulled. The hammer is released and the gun fired once the trigger is pulled far enough. For autoloading pistols, the self-loading mechanism will also recock the hammer after the first shot so that subsequent shots are fired single-action. Double-action-only (DAO) pistols do not use the motion of the slide to recock the hammer. Rather, both for the initial shot and all subsequent shots, the hammer is cocked by releasing the trigger.

Firing a gun

When a single or semi-automatic pistol is fired the hammer is cocked to draw back the slide. Pulling the trigger then discharges the cartridge. This causes the slide to recoil backwards. As the slide recoils it opens the breech, ejecting the casing. At the same time the slide recocks the hammer for the next shot. The spring-loaded slide mechanism automatically returns to its position and a new cartridge is fed into the magazine.

How the modern revolver works

A revolver has several firing chambers arranged in a circle in a cylindrical block. When the trigger of the revolver is pulled, the cartridge in the chamber is aligned with the barrel. At the same time the hammer is released and a compressed spring drives the hammer forward. The firing pin on the hammer extends through the body of the gun and hits the primer. The primer explodes, igniting the propellant. The propellant burns, releasing a large volume of gas, which drives the bullet down the barrel. The pressure of the gas also expands the cartridge case which has the effect of temporarily sealing the breech. At the same time the trigger is pulled, a mechanism attached to the trigger pushes on a ratchet to rotate the cylinder. This positions the next breech chamber in front of the gun barrel ready for firing.

ABOVE When the spent shells have been removed from the cylinder, individual cartridges can be placed in the chambers.

How a semi-automatic pistol works

A semi-automatic pistol works by extracting and ejecting a fired cartridge automatically from a chamber, which then loads an unfired cartridge from a magazine into the chamber to be ready for the next trigger pull. This cycle uses the energy of the explosive discharge of each cartridge that is fired.

1. A magazine is loaded with 9mm cartridges, and then placed into the pistol's grip. The slide is pulled back. This cocks the hammer and is then released and is pushed forward by the recoil spring. The breech block feeds the first round of ammunition from the magazine into the barrel's chamber. The next round in the magazine is moved into position.

2. Now the pistol is loaded, the trigger is pulled. This draws the trigger bar forward. This pivots the safety lever, raising the firing pin safety block so the firing pin can move. Simultaneously, the trigger bar pulls the sear forwards, releasing the hammer. The main spring pushes the hammer forwards so that it strikes the back of the firing pin. This hits the cartridge and the primer causes the gun to fire.

3. The force of driving the bullet out of the barrel also presses the empty cartridge case backwards as a result of recoil. The force also pushes the slide to the back. The firing pin safety block is activated by its separation from the safety lever. As it moves backwards the slide forces the trigger bar downwards and away from the safety lever. The slide also takes the barrel with it as it travels backward. The locking insert stops the backward motion and moves it down and away from the slide.

4. As the slide returns, the sear and safety lever go back to their starting points. The empty cartridge case is pushed upwards out of the ejection port. As the slide recoils to its stop, it cocks the hammer. The slide is pushed forward by the recoil spring, and the next round is fed into the chamber of the barrel as the slide locks up with the barrel again. The pistol can be fired again.

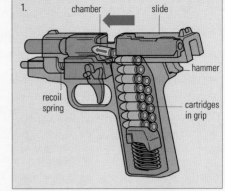

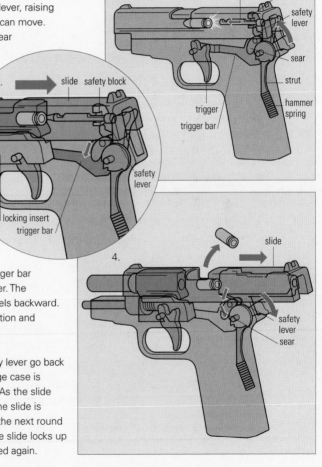

The 21st century and the future

Non-lethal hand-held guns are now regarded as useful weapons, whereas 20 years ago a handgun was used simply to kill an assailant. Research continues into how to produce weapons that will stop an aggressive person without killing or severely injuring them. On a larger scale, riots present a different but related challenge. Arms manufacturers have produced a variety of solutions, from incapacitating gases that generate coughing and disorientation, but not lasting injury, to rubber bullets and anaesthetizing darts.

CS gas and baton rounds

In 1928 two American scientists, Ben Carson and Roger Staughton, discovered 2-chlorobenzalmalononitrile, and the first two letters of the scientists' surnames were used to name the substance. This CS gas was developed and tested secretly at the British Chemical and Biological Warfare Establishment at Porton Down, Wiltshire, during the 1950s and 1960s. CS is actually a solid chemical and in order to deliver it to a human target it must be converted into a smoke or aerosol. In powder form it can be incorporated into

ABOVE Police firing rubber bullets and tear gas as they struggle to stop supporters of Venezuelan President Hugo Chavez from clashing with opposition protesters, Caracas, 4 November 2002.

riot-control munitions like baton rounds. The 12-gauge Ferret barricade round from Mace Security of the United States is designed to punch through barriers such as car windshields or wooden doors and scatter a quick-acting liquid irritant agent inside. However, unlike a baton round, the Ferret breaks up on impact with a barrier.

The British developed baton rounds in the late 1960s, when colonial police in Hong Kong used 51mm/2in wooden cylinders fired from large-bore weapons to disrupt riots. The aim was to replicate the effect of using a truncheon on demonstrators, without the police actually having to engage in hand-to-hand combat with the rioters.

The rounds were expected to fall a few feet short of rioters, to ricochet off the ground at about ankle height. However, the police found the wooden rounds often split, had unpredictable characteristics, and could prove lethal if fired at close range. When the "troubles" began in Northern Ireland in the 1970s, British troops were equipped with baton-round guns that fired tough rubber bullets, which were later replaced by PVC rounds. The use of baton rounds became increasingly controversial as injuries and deaths were linked to close-range hits.

Rubber bullets

The French company Manurhin, with a well-respected reputation for its small arms, devised the MR-35 Punch. This is a 35mm five-shot weapon that uses a small powder charge to fire a soft rubber ball. A hit will not kill, but it does deliver the sort of impact that you would feel if hit by a .38 Special bullet. Five balls can be fired from the MR-35 Punch in as many seconds and it can hit a $17cm^2$ target at 7m ($6in^2$ at 20ft). Made largely from polymers, the MR-35 weighs only 1,500g/53oz loaded.

A similar concept is reflected in the American RB rubber ball series of 12-gauge cartridges. These contain nine small hard rubber balls that are designed to ricochet off the ground. Again, this firearm will knock down, wind and deter a hostile individual without causing fatal injury.

The anaesthetic gun

The Chinese BBQ-901 anaesthetic gun system is a slightly sinister device for neutralizing an assailant and is probably derived from game conservancy dart rifles. It is a single-shot hand-held weapon that fires a projectile, that on impact, injects a liquid anaesthetic. Depending on the type of anaesthetic used, this will render the target unconscious in one to three minutes. The effects of the anaesthetic wear off in three to four minutes – which gives enough time for the subject to have been handcuffed.

The grenade launcher

The South African MGL Mark 1 40mm grenade launcher is a six-shot shoulder or hip-fired revolver that fires baton rounds or tear gas cartridges. It has a minimum range of 30m/98ft and maximum of 400m/1,312ft. Like gun fighters in the Wild West, a skilled operator can fire six rounds in just three seconds.

LEFT A Metropolitan Police firearms officer with a L104AL baton gun. This weapon can be deployed at a distance no closer than 1 metre from a suspect. The L104AL fires a high-velocity plastic round.

Knock down

One of the most effective non-lethal systems available to police and security services is the Taser, developed by Taser Inc. in the United States. The hand-held device fires two probes up to 5m/16ft carrying a lightweight wire. Once the probes have attached themselves to an assailant, an electrical signal is transmitted causing him to lose neuromuscular control. Once he has been knocked down, a series of pulses keeps him down and ensures that he cannot remove the probes. The Taser has become an ideal device for controlling aggressive or dangerous individuals. Unlike a CS spray that must be directed at the face, or a handgun that has to be fired at a head or torso, the Taser works on whatever part of the body it hits.

ABOVE British Police used a non-lethal Taser at Manchester Airport in September 2005 to immobilize a man who was acting suspiciously and resisting arrest.

Index

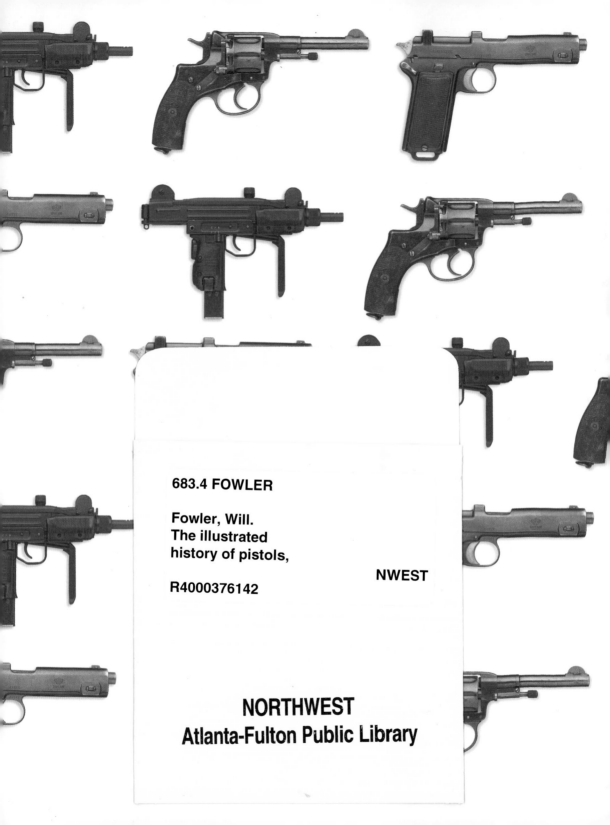